Tom Leaf

MANUAL FOR ENVIRONMENTAL IMPACT EVALUATION

MANUAL FOR ENVIRONMENTAL IMPACT EVALUATION

SHERMAN J. ROSEN

PRENTICE-HALL, INC.
Englewood Cliffs, New Jersey

Library of Congress Cataloging in Publication Data

ROSEN, SHERMAN J (date)
    Manual for environmental impact evaluation.

    Includes bibliographical references and index.
    1. Environmental impact statements. I. Title.
TD194.5.R67   363.6   75-20647

© 1976 by Prentice-Hall, Inc.
Englewood Cliffs, New Jersey

All rights reserved. No part of this book
may be reproduced in any form or by any means
without permission in writing from the publisher.

10 9 8 7 6 5 4 3

Printed in the United States of America

PRENTICE-HALL INTERNATIONAL, INC., *London*
PRENTICE-HALL OF AUSTRALIA, PTY. LTD., *Sydney*
PRENTICE-HALL OF CANADA, LTD., *Toronto*
PRENTICE-HALL OF INDIA PRIVATE LIMITED, *New Delhi*
PRENTICE-HALL OF JAPAN, INC., *Tokyo*
PRENTICE-HALL OF SOUTH EAST ASIA, PTE. LTD., *Singapore*

CONTENTS

	Preface	*vii*
1	Introduction	*1*
2	Background	*4*
3	History of Project Review	*10*
4	The National Environmental Policy Act of 1969	*12*
5	Public Participation and Agency Coordination	*31*
6	Assessment Concepts	*36*
7	Pollution Possibilities	*47*
8	Ecological and Environmental Inventory	*63*
9	Socio-Economics	*74*
10	Land Use	*79*
11	Historic and Archeological Sites	*82*
12	Graphic Aids — Remote Sensing	*85*
13	The Draft Environmental Statement	*91*
14	Public Meetings and Hearings	*96*
15	The Final Statement	*99*
16	Author's Note	*101*

Appendices

1	National Environmental Policy Act of 1969	*103*
2	Council on Environmental Quality Guidelines	*111*
3	Office of Management Circular A-95, Revised	*147*
4	National List of State Clearinghouses	*167*
5	Section 4(f)	*173*
6	Mathematical Evaluation System	*177*
7	Sample Draft Environmental Statement	*189*
	Index	*229*

PREFACE

The passage of the National Environmental Policy Act (NEPA) fostered the need for environmental evaluation or assessment.

Ground rules and guidance were almost non existent as only the intent of the law was firmly established, leaving such minor difficulties as interpretation, procedure, scope and format to overcome.

To fill this void representatives of almost every interest and discipline were recruited or volunteered to act as practitioners of the new mystic "science."

Since that time many of the more pressing problems have been brought under control and a semblance of order has started to solidify.

During that period a greater interest in the evaluation process was generated throughout the scientific world as well as in all levels of government. However, no base book was available to even acquaint those people to the intricacies of this fledgling field.

The need was truly dramatic; not so the literary efforts which required almost four years to finalize as philosophy, requirements, techniques and procedures were in a state of flux.

The need is still, in my opinion, dramatic; however, the manual is complete and hopefully will serve its intended purpose of introducing the evaluation process and some supporting rationale.

A full text on this subject matter with its broad range of environmental factors, relationships, and interrelationships will someday be published. However, in the interim I am hopeful this manual will serve as an introduction to the subject.

I wish to thank the following friends and associates for their time, efforts, and generous contributions without which this manual would never have been completed: Dr. Richard Wagner, Robert Blanco, Thomas Maurer, Joseph Layton, Prof. William King, Dr. David Long, Michael Oslac, Dr. Robert Scherer, Dr. Hayes Gamble, Dr. Ronald Shelton, and John V. Buffington, Attorney-at-Law, who wrote the chapter on NEPA Law. I also want to thank my doctoral committee: Dean Charles Olson, Prof. William Johnson, Dr. William Bentley, and Dr. Richard Andrews. Thanks also to my intrepid typist, Mrs. Doris DeHart; draftsman, Glenn E. Toberman; and finally to my wife Frances who spent many hours without her husband.

SHERMAN J. ROSEN

1

INTRODUCTION

And the Lord spake unto Moses, saying: "There is both good news and bad news. The good news is that plagues shall smite your Egyptian oppressors. The Nile shall be turned to blood, and frogs and locusts shall cover the fields, and gnats and flies shall infest the Pharaoh's people, and their cattle shall die and rot in the pastures, and hail and darkness shall visit punishment upon the land of Egypt! Then will I lead the children of Israel forth, parting the waters of the Red Sea so that they may cross, and thereafter strewing the desert with manna so that they may eat."

And Moses said, "O Lord, that's wonderful! But tell me, what's the bad news?"

And the Lord God replied, "It will be up to you, Moses, to write the environmental-impact statement." *

Section 102(2)(c) of the National Environmental Policy Act of 1969 [1] requires an environmental evaluation of the major federal actions that could significantly affect environmental quality. The review is known as the NEPA assessment process, and the resulting document is called an Environmental Impact Statement (EIS).

* Originally appeared in PLAYBOY Magazine; copyright © 1975 by Playboy.
[1] Appendix 1.

The stipulation includes projects or programs initiated by other levels of government and those in private industry, which require some form of federal involvement. This language extends the scope of actions requiring an EIS to an extremely large percentage of federal activities.

In addition, several state governments and regional and local agencies require independent environmental review. Many public utilities are also responsible to federal and state agencies in meeting this responsibility before project approval.

Although greater impact on physical environmental factors will generally result from major construction, on occasion other proposals also require adequate assessment so as not to reduce environmental value. In general, however, the more massive and complicated actions require more comprehensive EIS.

Statements are currently being prepared by individuals or teams from government agencies as well as by engineering and environmental consultants.

Note that the act did not identify which specific actions required statements except to stipulate that they would have the potential to significantly affect the environment. This lack of definition, further complicated by there being no definition for the word *significant*, has created some misunderstanding. These problems, however, have been and are being solved as those involved in the process can draw on court decisions and their own experience to establish definitions and parameters.

It has become apparent, however, that the number of projects or programs requiring this type of evaluation will increase and most likely eventually include all major construction, even projects financed by private sources. This is an important goal as it is both inconsistent and inefficient to rigidly control government-sponsored construction and allow privately-financed projects to proceed without comparable safeguards. As the requirement for environmental evaluation expands, it will become necessary for more people to become familiar with both procedures of preparation and review.

The NEPA program is beginning to mature, and guidelines have been reasonably well established, but changes continue to occur as a result of research, judicial interpretation, and the improved perspective resulting from experience. The number of environmental relationships is almost infinite and many newly charged with the responsibility of preparing and reviewing environmental assess-

Introduction

ments, are working on a trial and error basis and finding the task extremely difficult. This manual has been written to provide those not familiar with it a basic understanding of the NEPA process, its requirements, approaches to the preparation of a statement, and finally, a few suggestions to make their task a little easier and possibly a great deal more meaningful.

The latest Council on Environmental Quality Guidelines, which is considered the basic document regarding NEPA from that Agency, can be found in the Appendix.[2]

Suggested Reading

Thomas G. Dickert, editor, *Environmental Impact Assessment: Guidelines and Commentary* (University of California, Extension Press).

[2] Appendix 2—"Council on Environmental Quality Guidelines."

2

BACKGROUND

There is nothing new about concern for the environment. For years, professionals in the constituent disciplines of environmental study have warned of the serious problems we now face, but their warnings have been largely unheeded.

They have pointed out the results of our callous disregard for environmental quality and ineffective land management. In addition, they have explained danger signals of future problems and demonstrated the need for immediate action. Thanks to their persistent efforts, public concern has now been aroused and is resulting in demands for solutions to the problems besetting our environment.

This interest in our surroundings, the newfound appreciation for their value and concern for their potential loss, is certainly justified. Most forms of pollution have become widespread, requiring stringent, yet workable, controls. These must not be idealistic or impractical, but logical criteria that can be both adhered to and enforced. In addition, that land has been poorly used is obvious across the nation, and the need for rational land management criteria and subsequent efficient policy is critical.

This newfound public concern has become a catalyst for legis-

Background

lative and judicial action, one focal point of which is the controversial Environmental Impact Statement.

The *Statement* is an official document that is submitted by the federal or state agency proposing to undertake or fund an action or a project. The actual report, however, can be and often is prepared by local or state units, universities, or consultants. It is then developed with guidance from and reviewed and accepted by the agency responsible, before it is officially submitted.

The statement defines and evaluates the effects on the environment of a proposed project or action and its alternatives. It further attempts to determine the possibility of ameliorating negative impacts by creating favorable "trade offs."

Included in the alternatives of every action is the option of doing nothing (no-action). On construction projects non-construction solutions should also be addressed. If the environmental impacts prove to be sufficiently serious, the project may or may not be abandoned, depending on other considerations, such as need. For example, the oil shortage resulted in the recent approval by Congress of the construction of the Alaskan Pipeline, in spite of any environmental considerations.

The Statement must be placed in its proper perspective. It is a tool prepared to assist the decision maker in making sound and rational decisions regarding the environmental effects of various alternatives. The Statement itself does not make that decision; it is only part of the data that are used to make a final determination. Other factors include need, cost, public benefit, and the possibility of timely implementation. As the process is refined, statement and assessment data will also serve in the planning process, not only for specific projects, but also in the development of regional and local land use policies.

Unfortunately, the EIS program suffers from a widespread lack of understanding. It has been depicted as burdensome, costly, and time consuming. There is no question that all these effects have been encountered; however, the process has just left the formative stages, and the responsibility for such accusations has fallen as often on the shoulders of those preparing the statement as on the system itself.

Much of this has been due to the fact that requirements, in terms of content, depth of analysis, and format, were not explicit, which resulted in widespread confusion. This condition, however has been greatly improved recently to the extent that there is a fuller understanding of the requirements by those engaged in preparation and review.

Laymen and professionals have had in common the lack of a uniform interpretation of how the EIS process should operate. The controversy has not only existed between those who might exploit our natural resources and those who want to protect them; it also has divided the ranks of the most sincere conservationists and environmentalists. As in most emerging movements, success has spurred as many problems as solutions. Although experience has offered a better understanding of practicality, there is still disagreement as to approach, method, and procedure, which has contributed significantly to the environmental controversy.

During this formative period the pendulum has swung too hard. The result has been the creation of emotional environmentalism, which amounts to concern supported by sentiment, rather than scientific knowledge. This phenomenon is sufficiently widespread so as to be recognized as a practical facet of environmental evaluation that cannot be ignored, particularly in projects with political overtones.

Most levels of government have been and are proposing, debating, and enacting environmental legislation since the issue enters the political arena. The federal government, several states, counties, and cities have created agencies charged with the protection of the environment. Legal staffs set up to prosecute violators have shown spectacular enthusiasm and results.

A member of the Cabinet was summarily dismissed, at least partially because of environmental considerations. The bench and bar have been reviewing the legal ramifications of environmental legislation. However, an overwhelming proportion of early judicial decisions set precedent due to the newness of environmental actions. It is only recently that a clear-cut consistent line of logic has been established.

Citizens banded together to stave off environmental degradation in groups ranging from small associations fighting local issues to well-financed, ably-represented national organizations.

On January 1, 1970 the "National Environmental Policy Act of 1969" (NEPA)[1] was passed. The bill set forth a national environ-

---

[1] Public Law 91-190, 91st Congress, Chapter 4 of Text.

mental policy and established the Council on Environmental Quality to assist and advise the President on environmental and ecological matters.

Passage of this act was a major victory in the battle to conserve natural resources and protect the environment. It forcefully demonstrated the federal government's concern for the gravity of the problem, formalized its interest in establishing necessary policy and programs to help solve it, and established the mechanics needed to preserve and enhance environmental quality in the nation and its territories.

NEPA states, "The Congress authorizes and directs that to the fullest extent possible: all agencies of the Federal Government shall include in every recommendation or report on proposals for legislation and other major Federal actions significantly affecting the quality of the human environment, a detailed statement by the responsible official."

This, then, is the mandate for the Environmental Impact Statement. It is required as part of all major federal actions, direct and indirect, that might significantly affect the environment.

Those actions include, but are not restricted to, in addition to construction and maintenance functions, i.e., dredging, such functions as:

1. Funding of projects proposed by other levels of government and private organizations.
2. Issuing of permits by responsible agencies for changes or improvements within their jurisdiction.
3. The leasing of federal land or resource rights.
4. Legislation
5. Administrative duties such as approval for line abandonment by the Interstate Commerce Commission.

The machinery required to make this program operable has proved to be extensive at both federal and state levels. At this time many agencies have been organized or restructured to carry the additional responsibilities they have been assigned. As previously mentioned, guidelines and procedures, although generally complete, are still being updated to expedite the process and supporting paper work. Slowly, a semblance of order has emerged.

Standards are still being developed, however, basic requirements have been generally resolved, particularly in the pressing problems relating directly to air, water, and noise pollution. It is hoped that in the near future, nationally accepted standards, guidelines, and evaluation procedures will be established to insure uniformity of protection.

Even now most agencies suffer from a shortage of personnel and funding devoted to environmental evaluation. These conditions have left the work in the hands of an overloaded few, which has slowed down the NEPA process.

By interpretation the Environmental Impact Statement is the responsibility of the federal agency. However, Statements are being prepared by or under the administration of state or local agencies [2] and/or privately owned firms.[3] This allows for the use of outside practitioners when work loads preclude meeting of schedules or when specialized expertise is required. This is important as agency personnel is limited both in number and in geographic location which limits flexibility and familiarity with local situations.

Long distance evaluation is generally not as comprehensive, sensitive, or effective as local efforts. Future actions will require environmental analysis in perspective with these actions, as it has been proven that anything less could result in delays and increased costs.

An EIS must be approached with care, caution, and an open mind. It cannot be used as a "whitewash" instrument; nor, any longer, can unavoidable negative impacts be hidden or ignored. The program has progressed beyond that phase.

As a document, the statement should clearly define the effects, both positive and negative, the alternatives and their associated effects, and the measures proposed to alleviate those impacts. Under some conditions a potentially dangerous project may be adjusted so as to improve rather than reduce environmental quality.

[2] In fact the Community Development Act of 1974, Public Law 90-838, 93rd Congress (Aug. 22, 1974) has just recently placed more responsibility on local governments and even requires local officials to act as agents of the federal government in court actions.

[3] Even more recently, however, in the "see-saw," the Federal Second Circuit Court of Appeals upheld a decision that stated, among other findings, that the Federal Highway Administration must prepare its own Environmental Impact Statements in order to comply with NEPA. This decision has acted as the basis for the introduction of a congressional bill specifying that Federal agencies, though ultimately responsible, may delegate the preparation efforts to state and/or other local agencies.

Background

An informative statement is the result of comprehensive and logical assessment. The field of environmental protection is inducive of so many disciplines that no one man's knowledge can encompass its entire spectrum. Therefore, depending on the complexity of the project, an interdisciplinary approach must be considered as the most efficient method of determining potential impact.

3

HISTORY OF PROJECT REVIEW

Before the National Environmental Policy Act of 1969, and as far back as 1966, the federal government was developing the concept of considering project impact on local planning goals. It was based on the premise that local assistance and influence could be brought into play by a system of project notification that would alert associated agencies and government bodies to a proposed project and offer an opportunity to express their opinion as to compatibility with existing plans. This was an attempt to bring some systematic approach to the projects the various departments were supporting by virtue of loans and grants.

Section 204 of the Demonstration Cities and Metropolitan Development Act of 1966[1] set up this system of project notification and review on a regional basis. The program was generally successful except for a few cumbersome procedures.

The Inter-governmental Cooperation Act of 1968[1] was enacted to further develop and sophisticate the system of notifica-

---

[1] Attachments A and B, Circular A-95, Appendix 3.

tion and review, offering broader and more efficient procedures evaluation.

The Office of Management and Budget (OMB) established the procedures in Circular A-80, dated January 31, 1967. Circular A-82 Revised, which was issued January 10, 1969, updated those procedures.

On July 24, 1969, OMB issued Circular A-95 with regard to evaluation, review, and coordination of federal assistance programs. This circular further streamlined procedures so as to minimize delay and expand benefits.

With the passage of the National Environmental Policy Act of 1969, including the EIS requirement of Section 102(2)(c) and the establishment of the Council on Environmental Quality, the bureau again revised the Circular A-95 on February 9, 1971. These procedures became effective on April 1, 1971.[2] A-95 now defines the current framework of operation and the affected programs. The circular has been printed *in toto,* resulting in some repetition, to avoid taking any part of this important document out of context.

Notification must be given for proposed projects by a letter of intent addressed to the state and regional clearinghouses,[3] and at the same time environmental impact evaluation should be in process.

[2] Appendix, Council of Environmental Quality Guidelines—Appendix 2.
[3] Appendix, National List of State Clearinghouses—Appendix 4.

4

THE NATIONAL ENVIRONMENTAL POLICY ACT OF 1969

As the cornerstone of environmental legislation, NEPA, which was passed into law on January 1, 1970, is included in its entirety in Appendix 1. The four key sections of the act involve:

1. The declaration of a national policy and the intent of implementing actions that will serve to both improve and protect the national environment.
2. The establishment of the Council of Environmental Quality to serve and advise the President in environmental matters.
3. Section 102(2)(c), which requires a detailed statement on "proposals for legislation and other federal actions significantly affecting the quality of the 'environment'" on:

(i) The environmental impact of the proposed action;
(ii) Any adverse environmental effects which cannot be avoided should the proposal be implemented;
(iii) Alternatives to the proposed action;
(iv) The relationship between local short-term uses of man's environment and the maintenance and enhancement of long-term productivity; and

(v) Any irreversible and irretrievable commitments of resources that would be involved in the proposed action, should it be implemented.

4. Section 102(2)(d) which requires those agencies responsible to:

Study, develop, and describe appropriate alternatives to recommended courses of action in any proposal which involves unresolved conflicts concerning alternative uses of available resources.

Currently, almost all federal agencies have prepared at least interim, and in many cases final, guidelines delineating their EIS procedures for complying with NEPA.

Directly related to the establishment of agency concepts have been a series of judicial interpretations which have been largely responsible for shaping the NEPA process and its procedures, particularly in relation to Section 102(2)(c).

In writing this manual I discovered that a meaningful description of NEPA's judicial history would be extremely difficult for a layman; therefore, the following section has been prepared by an attorney who is vocationally and avocationally interested in environmental law.

His text has been printed as submitted and although the format differs slightly from the rest of the book, I feel it would be an injustice to tamper with this important contribution, which points out the bases of legal argument, history of important legal decisions, and finally, a concise insight into those decisions.

• • •

JUDICIAL INTERPRETATIONS OF NEPA
JOHN V. BUFFINGTON Attorney at Law

This article is intended to be a summary and synthesis of the holdings to date of the federal courts concerning the environmental impact statement process. The holdings which are rather baldly stated herein are the products of extended analysis in specific factual settings. Consequently, reference to the rules set forth cannot substitute for conscientious effort on the part of all concerned to make environmental decision-making responsible.

    Philadelphia   *January 24, 1975*

Footnotes for "Judicial Interpretations of NEPA" begin on text page 26.

The courts have grappled repeatedly with the question of what constitutes a "major federal action significantly affecting the quality of the human environment." Most are agreed that two separate inquiries must be made: whether the action is "federal," and whether it will have a "significant effect" on the environment.

"MAJOR FEDERAL ACTION"

Whether a project is federal or not depends upon the level of federal agency involvement. Where there is no federal involvement and no federal funds are to be utilized, the action is not federal and NEPA is inapplicable.[1] Federal involvement can be shown where the federal government is involved in the promotion or planning of the project.[2]

An action is federal if it will affect the status of federal lands, such as the approval of leases[3] or the exchange of federal land for privately owned land.[4] Federal involvement in the promotion or planning of the project makes it federal.[5] Federally funded research programs can be major federal actions.[6] In fact, the commitment of large amounts of federal funds will make any project federal, regardless of the level of federal control.[7] The Ninth Circuit may be in conflict with the majority view on this point, having held the payment of farm subsidies not to be a major federal action because the program was mandatory and the government had no control over how the money was spent.[8]

On the other hand, the circuits are agreed that certain kinds of minimal federal participation in projects will not make them federal. Comment by a federal agency on a project, which it has no statutory obligation to approve, is such a case.[9] Federal Power Commission approval of expansion of an existing facility was held not to be a federal action by the Third Circuit.[10] Corps of engineers issuance of dredging permits may not be a federal action under NEPA.[11]

For the most part, the courts have ignored the question of what makes a federal project "major," tacitly assuming what the Eighth Circut has made explicit: that the magnitude of a federal action cannot be separated from the significance of the effect on the human environment. If the action has a significant effect, NEPA necessarily applies.[12]

The Second Circuit disagrees, however, and has enunciated what is probably the better rule: that whether a federal action is "major" or not depends upon the cost of the project, the amount of planning that preceded it, and the time required to complete the project, rather than the impact of the project on the environment.[13]

EXCLUSIONS

While in most cases the question of whether an action is federal or not turns upon the level of agency involvement, a question sometimes arises as to whether the agency itself is subject to NEPA. This has been a problem with regard to agencies of local government in the District of Columbia. The D.C. Redevelopment Land Agency must comply[14] and other D.C. government agencies may be bound by NEPA, since they are responsible to Congress.[15] On the other hand, the D.C. City Council, which is popularly elected, is not a federal agency.[16] The D.C. Zoning Commission is a local agency, while the National Capital Planning Commission is federal.[17]

Although NEPA applies to federal activities anywhere under the control of the United States, and even to actions affecting the environment in other countries,[18] the Ninth Circuit has held that the government of the Pacific Islands Trust Territory is not a federal agency subject to NEPA.[19] The court reached this conclusion because governments of territories and possessions are excluded from the Adminstrative Procedure Act. This argument is somewhat unpersuasive, in light of the District of Columbia cases mentioned above, and the broad policies stated in NEPA.

Another exception to the applicability of NEPA has been delineated by the Tenth Circuit, which has declared that the traditional unfettered control of the internal management and operation of federal military establishments is not altered by the Act. Where matters are committed wholly to official discretion, no impact statement need be done. The same is true of matters touching national security.[20]

The courts have inferred a Congressional intent to exclude from the application of NEPA administrative actions taken by agencies engaged primarily in an examination of environmental issues, provided that appropriate safeguards are employed to insure that NEPA's policies will be fulfilled.[21]

While Congress clearly has the power to partially repeal NEPA by including exclusions in subsequently passed legislation, and has done so, for example, in Section 511(c) of the Federal Water Pollution Control Act Amendments of 1972, the courts are most reluctant to presume such a partial repeal, absent very clear statutory language setting it forth. Passage of an appropriations bill funding a contested project is not such a partial repeal, without a specific statement of that intent,[22] nor is Congressional requirement of special clearance of a project by a department head[23] or the President.[24]

The courts will, however, infer Congressional intent to partially repeal NEPA where emergency administrative actions are required, and the time periods specified are obviously too short for procedural compliance.[25]

Where time periods in legislation passed prior to NEPA are too short for full compliance, the agency is not required to follow the NEPA procedures, by virtue of the inclusion in Section 102 of the phrase "to the fullest extent possible."[26] This exception, however, applies only where there is a clear *statutory* conflict.[27]

"SIGNIFICANTLY AFFECTING"

One court has defined the phrase "significantly affecting the quality of the human environment" as "having an important or meaningful effect, direct or indirect, upon a broad range of aspects of the human environment."[28]

That definition is probably too narrow, as is evidenced by decisions of a number of courts, which conclude that some single effect is or is not significant. For example, taking ten acres of parkland in a thickly settled city has a significant effect,[29] as does the discharge of heated water into a river,[30] or the working of changes in the architectural character of a neighborhood, accompanied by the addition of more residents.[31]

On the other hand, the introduction of a public housing project into a neighborhood made up of persons of different socio-economic characteristics does not constitute a significant effect;[32] nor does the condemnation of land and cutting of trees for a power line.[33]

Where the federal government adheres to local zoning regulations, it is evidence and may create a presumption that the project's impact on the environment is not significant.[34]

Finally, it is worth noting that the Fifth Circuit has held that a significant effect can be a positive effect; the claim that all effects of a project will be beneficial will not serve to excuse compliance with NEPA.[35]

NEGATIVE DECLARATIONS

A federal agency which is contemplating taking an action has the authority to determine whether there is sufficient probability of the action's having a significant effect on the environment to require an impact statement.[36] No statement is necessary in cases of "true insignificance," but cases of "arguable significance" should be resolved in favor of preparing the statement.[37]

The courts tend to give considerable deference to the determination of the responsible agency. The Second and Seventh Circuits will reverse only upon a showing that the decision was arbitrary, capricious, or an abuse of discretion,[38] while the Tenth Circuit finds that NEPA sets a very high standard for agency conduct which the courts must enforce.[39] The Fifth Circuit will also make its own determination as to whether the agency made the correct decision.[40]

Whatever standard of court review is followed, the sponsoring agency must follow procedures which insure that it reaches a fair and informed preliminary decision as to whether an impact statement is required,[41] and must provide a statement of reasons in the event that its findings are negative.[42] Such a statement must be well reasoned, and cannot simply be couched in perfunctory and conclusory language.[43]

The Second Circuit holds that agency procedures must provide for examination of at least the following:

1. The extent to which the action will cause adverse environmental effects in excess of those created by existing uses in the area affected by it;
2. The absolute quantitative adverse environmental effects of the actions itself, including the cumulative harm that results from its contribution to existing adverse conditions or uses in the affected area.[44]

The District of Columbia Circuit's standard for agency procedures, on the other hand, requires the following:

    1. A hard look at the problem, as opposed to coming to bald conclusions;
    2. Identification of the relevant areas of environmental concern;
    3. A convincing explanation as to why the impact is insignificant;
    4. Any necessary changes in the project to minimize any significant impact.[45]

RESPONSIBILITY FOR PREPARATION

Where two or more federal agencies are involved, a decision must be made as to whether separate statements will be prepared on each agency's portion of the project, a joint statement will be prepared, or one agency will take primary responsibility for preparation of a single statement on all aspects of the project. Where the latter course is adapted, the agency responsible for impact statement preparation is called the lead agency.

The two cases in which the courts have addressed the question of which should be the lead agency have indicated that the agency owning the land and supervising the work takes precedence over the one funding the project, or one that will occupy it.[46] Both also uphold the determination of the agencies involved. A third case holds that such an agreement between agencies is a purely administrative matter that is not reviewable by the courts.[47]

DELEGATION

The Second Circuit held in January, 1972, that the federal agency responsible for taking the action must prepare the impact statement itself, and cannot delegate this responsibility to a state agency or other interested party.[48] This is still good law in the Second Circuit[49] but the other circuits have declined to follow it.

The majority rule now is that a federal agency can delegate the task of impact statement preparation to a state agency provided that the federal agency reviews the statement and adopts it as its own. This is more or less the stance adopted by the Eighth, Ninth and Tenth Circuits[50] and is probably acceptable to the Fifth Circuit.[51]

The Fifth and Ninth Circuits have also held that a private company with a financial interest in the project can draft the impact statement.[52] The federal agency must, however, play a significant role in the preparation of the draft, since the private company will clearly have a conflict of interest in performing this responsibility. This conflict of interest problem may very well be sufficient to prevent the acceptance of this rule by other circuits.

INNOCENT BYSTANDERS

Where federal agencies fail to properly discharge their responsibilities under NEPA, other parties involved in the project may well be hurt. The Fourth Circuit has held that a state cannot be enjoined with the federal agency,[53] but that the funds involved must be returned to the U.S. Treasury, not used on another project.[54] Other circuits may well be willing to enjoin state agencies directly.[55] Private parties may be enjoined as well, if they act in partnership with a federal agency that fails to comply with NEPA.[56]

Even if not enjoined, a state agency or private party proceeds at its own financial risk with respect to any investment made prior to the completion of judicial review of NEPA compliance.[57] The wisest thing is always to avoid altering the status quo until the NEPA process is clearly complete.[58]

CONTENT OF THE IMPACT STATEMENT

The basic elements of the impact statement are listed in Section 102(2)(C) of NEPA. Each of these elements must be adequately treated in order for the statement as a whole to be considered sufficient.

The proposed action. The impact statement must describe the proposed action to be taken,[59] giving sufficient detail to enable persons unfamiliar with the project to understand and meaningfully consider the factors involved.[60]

Impact of the proposed action. The impact statement must deal with all significant effects on the environment which are reasonably anticipated. Projected increases in noise, traffic and the burden on mass transit must be treated,[61] as must any other expected results of population and industrial growth that may be triggered by the project.⁶² Sufficient detail should be provided to enable a responsible executive to arrive at a reasonably accurate decision regarding the environmental benefits and detriments to be expected from program implementation.⁶³

Alternatives. The alternatives section of the impact statement must discuss all alternatives which are reasonably available, but need not deal with those which are speculative and remote.⁶⁴ One such choice is no action or total abandonment,⁶⁵ although in the case of ongoing projects, the previous investment in the project as proposed may be taken into account in evaluating the possibility of abandonment.⁶⁶ Reduction in the scale of the project must also be considered.⁶⁷

Alternatives which do not offer a complete solution to the problem must be considered.⁶⁸ The sponsoring agency must also examine alternatives which may be outside its jurisdiction or control,⁶⁹ including those which would constitute separate projects requiring separate Congressional authorization.⁷⁰

The statement must discuss alternatives suggested by responsible critics, and any action which might be taken in mitigation of damages.⁷¹

Section 102(2)(D) requires the sponsoring agency to "study, develop, and describe appropriate alternatives to recommended courses of action in any proposal which involves unresolved conflicts concerning alternative uses of available resources." The alternatives section of the impact statement is the appropriate setting for this discussion.⁷²

The alternatives which are discussed must be sufficiently detailed to show the basis for the decision-maker's conclusions.⁷³ The treatment need not always be exhaustive, but must be sufficient to allow a reasoned choice.⁷⁴

QUANTIFICATION

Section 102(2)(B) of NEPA requires agencies to "identify and develop methods and procedures . . . which will insure that presently unquantified environmental amenities and values may be given appropriate consideration in decision-making along with economic and technical considerations."

This passage had little or no effect upon the actual operation of government agencies, until the District of Columbia Circuit concluded in the *Calvert Cliffs'* case that it mandates a finely tuned and systematic weighing process which must be incorporated into the impact statement.[75] Where normal agency procedure requires calculation of the cost-benefit ratio for the project in the course of deciding whether to go forward, that analysis must reflect environmental costs and be incorporated into the impact statement.[76]

Moreover, in such a case, the statement must contain an explanation of how the benefits and costs were calculated.[77] Failure to give sufficient consideration to environmental values in computing the cost-benefit ratio is a basis for judicial intervention,[78] but the courts will probably allow the agency more latitude in its exercise of judgment here than in compliance with NEPA's procedural requirements.[79]

The Fifth Circuit appears to be at variance with this line of cases, holding that NEPA does not require actual quantification of environmental amenities, but only that the agency identify and develop methods and procedures reasonably calculated to bring environmental factors to peer status with dollars and technology in decision-making.[80]

METHODOLOGY

NEPA requires that each agency undertake the research necessary to adequately expose environmental damage that may result from the project.[81] The choice of which studies to make, however, is within agency discretion, and the courts will not interfere in the absence of a clear error of judgment.[82] The Ninth Circuit even holds that the Act does not require that the agency wait until all studies are complete before taking action.[83]

The agency must approach its consideration of environmental consequences in a spirit of "good faith objectivity." [84] This objectivity must be maintained all the way through the decision-making process.[85] The style of the impact statement should be objective as well, and combative, defensive, and advocatory language may be considered evidence of a lack of fairness in the agency's consideration.[86]

The decision-makers must consider environmental factors discussed in the statement throughout the agency review process.[87]

Furthermore, once the process is complete, the agency must allow sufficient time to consider the finished statement before taking action. Rubber stamping of the recommendations of the staff will not suffice.[88]

TIMING

The impact statement must be prepared early in the administrative process leading to approval of a project.[89] It should be done prior to any formal hearings on the project.[90] The courts are reluctant, however, to find that proper timing has not been followed, and tend to conclude that the agency complied with the Act sufficiently, barring other deficiencies.[91] Three circuits have even refused to enjoin projects in which they specifically found that the statements were prepared rather too late in the process.[92]

This curious refusal to provide a remedy for a recognized abuse may reflect peculiarities in the fact situations in which the timing question has come up. Some courts have shown a willingness to enjoin projects in which proper timing has not been followed,[93] and it seems likely that the circuits will become more insistent upon early impact statement preparation in response to increased pressure from the plaintiff bar.

CONSULTATION WITH OTHER AGENCIES

Section 102(2)(C) provides that in the course of impact statement preparation, the sponsoring agency must "consult with and obtain the comments of any federal agency which has jurisdiction by law or special expertise with respect to any environmental impact involved."

One court has held that the purpose of this requirement is to avoid obvious duplication of time and expense,[94] whereas another has stated that it is a way of insuring that the required analysis was undertaken.[95] In any event, the coordination is mandatory, and where changes are made in a project after reviewing agencies have commented, further coordination and review must take place.[96]

Although the sponsoring agency must give considerable deference to the views of other agencies commenting within the proper scope of their review,[97] it may not rely upon favorable findings by other agencies to the extent of failing to conduct the required case-by-case balancing of environmental and other factors.[98]

Any report by a federal agency recommending against the project, because of potential harm to the environment, must be obtained by the sponsoring agency and released to the public.[99] Comments submitted by other federal agencies or by the public must be included in the final impact statement,[100] and the text of the statement must draw the attention of readers to opposing views.[101] Submissions by other agencies which have nothing to say on the particular project, however, may be omitted.[102]

The District of Columbia Circuit is more permissive than the majority in this regard, and has held that the sponsoring agency has the discretion to exclude opposing views that are not responsible, and to discuss disagreements in the text of the statement, rather than incorporating the comments at full length.[103]

THE COUNCIL ON ENVIRONMENTAL QUALITY

Section 102(2)(B) gives the Council on Environmental Quality (CEQ) a role in agency development of methods for the quantification of environmental amenities. Section 102(2)(C) gives CEQ a role in reviewing impact statements. Thus it appears that Congress intended CEQ to play a role in supervising compliance with NEPA's requirements by the various federal agencies.

Accordingly, CEQ has issued Guidelines for federal agencies to utilize in implementing NEPA. The courts are in general agreement that these guidelines are merely advisory, since CEQ's mandate in NEPA does not include prescribing regulations governing compliance.[104] The guidelines are accorded great weight, however, and agencies whose procedures are not in compliance run grave risks of reversal in the courts.[105]

Moreover, in exercising responsibility to review statements, CEQ clearly has the authority to find that a statement is inadequate, and such a finding will in itself be accorded great weight.[106]

PUBLIC PARTICIPATION

The only mention of the public in Section 102(2)(C) is the requirement that the final impact statement be made available to the President, CEQ, and the public. Yet NEPA has worked a thoroughgoing revolution in the accessibility of the federal decision-making process to citizens and environmental organizations. The principal reasons for the change are (1) the increased visibility of the decision-making process; (2) the provision of new opportunities for public comment on proposals; and (3) the broadened availability of judicial review of the administrative process.

Administrative visibility. Before a preliminary determination of significance or negative declaration is made, the responsible agency must give notice to the public of the proposed action, and provide an opportunity to submit facts which might bear on the agency's initial decision.[107] The agency must circulate the draft impact statement to the CEQ and the public for comment in the course of the NEPA process.[108] When the statement is completed, the public must be given notice of its availability.[109] Critical agency comments must also be made available.[110]

Opportunity to comment. NEPA does not obligate agencies to hold public hearings on proposed projects, either in the course of making the preliminary determination of significance [111] or in the course of preparing a final impact statement.[112] But opportunities for submitting written comments must be provided at several key points in the process, and comments submitted must be fully considered by the sponsoring agency.[113]

Judicial review. NEPA has greatly expanded the availability of judicial review of federal decisionmaking. The courts have concluded that the only way to insure compliance with the policies and procedures established by NEPA is to allow citizens and environmental groups to act as "private attorneys general" [114] and sue to vindicate the national policies. Accordingly, the courts have inferred from NEPA an expansion of court jurisdiction,[115] have virtually eliminated the standing requirement as applied to environmental groups,[116] have progressively expanded the scope of judicial review of procedural compliance,[117] and have begun to review the substantive decisions reached by federal officials on the merits.[118]

SEGMENTATION

It is frequently possible for a federal agency to evade its environmental responsibilities by dividing its work up into numerous small projects, each of which individually has minimal environmental consequences, even though the federal program of which they are parts, taken as a whole, may have quite serious environmental consequences. The prevalence of this practice, especially in the realm of highway construction, has given rise to a large number of cases in which the courts have attempted to determine the limits of agency discretion in defining the size of projects to be evaluated.

Several cases have held that in the particular fact situations presented, environmental analysis must be performed on larger projects than were proposed by the sponsoring agencies.[119] In another, the court held that while no overall impact statement was required for a "program of projects" under the Federal Aid Highway Act, the sponsoring agency should consider the impact of the whole program in conjunction with one or more of the individual impact statements.[120]

The Eighth Circuit has held that a whole state highway plan need not be treated as a single unit for impact statement purposes. The segments that are utilized should be as large as possible, and supportable by logical termini at each end.[121] The First Circuit has held that projects which are separate for funding purposes need not be treated as one under NEPA.[122]

One court has held that dividing projects into segments is proper where it is based upon geographical differences and different highways needs, provided that the separate units serve separate, rational needs in their own rights; and no attempt is made to benefit the entire project by federal funds while avoiding federal statutes.[123] Another emphasizes that the length selected must assure adequate opportunity for consideration of alternatives, as to both whether to build and where to locate the route.[124]

Section 1653(f), Title 49, United States Code 1/, Section 138, Title 23, commonly referred to as Section 4(f) has also become an important provision for environmental control.

Public park lands, up until recently, have been taken for highway projects because of financial savings, expediency, and the lack of owner opposition. This section has radically changed that condition and demands proof that there is no "prudent or feasible alternative" to the taking of public land, that proper trade-offs will be offered, and that all steps possible to minimize destruction will be taken, before approval can be given.

Because of its importance, the text and requirements are included in Appendix 5.

The provisions obviously intended to protect public land have generally served as a safeguard and in the more populous metropolitan areas, that protection is extremely important. In fact, it has become recognized that more, not less, public land is required and many state and local governments are actively setting up funds for that purpose.

Footnotes for Judicial Interpretation of NEPA

[1] Bradford Township v Highway Authority, 4 ERC 1301, 1303; 463 F.2d 537; 2 ELR 20322 (7th Cir. 6/22/72). Cert. den. 4 ERC 1784 (12/5/72).

[2] Ely v Velde (I), 3 ERC 1280, 1285 (n.22); 451 F.2d 1130; 1 ELR 20612 (4th Cir. 11/8/71).

[3] Davis v Morton, 4 ERC 1735, 1738; 469 F.2d 593; 2 ELR 20758 (10th Cir. 11/24/72).

[4] National Forest Preservation Group v Butz, 5 ERC 1863, 1865 (9th Cir. 9/10/73).

[5] Ely I, p. 1285.

[6] Scientists' Institute v AEC, 5 ERC 1418, 1425; 481 F.2d 1079 (D.C.Cir. 6/12/73).

[7] Monroe County Conservation Council v Volpe, 4 ERC 1886, 1888; 472 F.2d 693; 3 ELR 20006 (2d Cir. 12/18/72).

[8] Kings County Association v Hardin, 5 ERC 1383, 1384; 3 ELR 20335 (9th Cir. 4/16/73).

[9] San Angelo Conservation Society v Texas, 6 ERC 1881, 1885; 496 F.2d 1017 (5th Cir. 7/5/74).

[10] Transcontinental Gas v Development Commission, 4 ERC 1441, 1447; 464 F.2d 1358; 2 ELR 20495 (3d Cir. 8/2/72). Cert. den. 4 ERC 2040 (1/8/73).

[11] Rucker v Willis, 5 ERC 1817, 1820 (4th Cir. 8/27/73).

[12] Minnesota PIRG v Butz, 6 ERC 1694, 1699 (8th Cir. 6/10/74).

[13] Hanly v Mitchell, 4 ERC 1152, 1155; 460 F.2d 640; 2 ELR 20216 (2d Cir. 5/17/72). Cert. den. 409 U.S. 990; 4 ERC 1745 (11/6/72).

[14] Jones v D.C. Redevelopment Land Agency, 6 ERC 1534 (D.C. Cir. 4/26/74).

[15] Citizens Association v Zoning Commission, 4 ERC 2063, 2068 (D.C.Cir. 2/6/73).

[16] Metropolitan Washington Coalition v Department, 5 ERC 1910, 1913 (D.D.C. 6/21/73).

[17] Tolman Laundry v Washington, 6 ERC 1264, 1272 (D.C. Sup. Ct. 2/1/74).

[18] Wilderness Society v Morton (I), 4 ERC 1101; 463 F.2d 1261; 2 ELR 20250 (D.C. Cir 5/11/72).

[19] Saipan v Department of Interior, 6 ERC 1952, 1954 (9th Cir. 7/16/74).

[20] McQueary v Laird, 3 ERC 1184, 1187; 449 F.2d 608; 1 ELR 20607 (10th Cir. 10/21/71).

[21] Environmental Defense Fund v EPA, 6 ERC 1112, 1119 (D.C.Cir. 12/13/73); Portland Cement v Ruckelshaus, 5 ERC 1593, 1600 (D.C. Cir 6/29/73); Essex Chemical v Ruckelshaus, 5 ERC 1820, 1823 (D.C. Cir. 9/10/73).

22 Environmental Defense Fund v Froehlke, 4 ERC 1829, 1834; 478 F.2d 346; 3 ELR 20001 (8th Cir. 12/14/72).

23 Davis, p. 1738.

24 Committee for Nuclear Responsibility v Schlesinger, 3 ERC 1276; 404 U.S. 917; 1 ELR 20534 (11/6/71).

25 Dry Color Manufacturers' Association v Brennan, 5 ERC 1961, 1967 (3d Cir. 10/4/73); Gulf Oil v Simon, 6 ERC 1966, 1968 (U.S. Temp. Ct. of Appeals 7/29/74).

26 Alabama Gas v FPC, 5 ERC 1010, 1016; 476 F.2d 142 (5th Cir. 2/7/73).

27 Alabama Gas, p. 1015; Monroe County, p. 1889; EDF v TVA, 4 ERC 1850, 1856; 468 F.2d 1164; 2 ELR 20726 (6th Cir. 12/13/72); EDF v Corps of Engineers (II), 6 ERC 1513, 1517 (5th Cir. 4/19/74); Calvert Cliffs' v AEC, 2 ERC 1779, 1782; 449 F.2d 1109; 1 ELR 20346 (D.C. Cir. 7/23/71). Cert. den. 404 U.S. 942 (1972).

28 Natural Resources Defense Council v Grant, 3 ERC 1883, 1890; 341 F. Supp. 356; 2 ELR 20185 (E.D.N.C. 3/15/72).

29 Monroe County, p. 1888.

30 Izaac Walton League v Schlesinger, 3 ERC 1453, 1454; 337 F.Supp 287; 2 ELR 20039 (D.D.C. 12/13/71).

31 Goose Hollow Foothills League v Romney, 3 ERC 1087, 1088; 334 F.Supp. 877; 1 ELR 20492 (D.Ore. 9/9/71).

32 Nucleus of Chicago Homeowners v Lynn, 6 ERC 1094, 1096 (N.D.Ill. 11/21/73).

33 Mowry v Central Electric Power, 5 ERC 1978, 1982 (D.S.C. 8/1/73).

34 Maryland Planning Commission v Postal Service, 5 ERC 1719, 1724 (D.C. Cir. 8/23/73).

35 Hiram Clarke Civic Club v Lynn, 5 ERC 1177, 1180; 3 ELR 20287 (5th Cir. 4/3/73).

36 Hanly v Mitchell, p. 1155.

37 Maryland Planning Commission, p. 1726.

38 Morningside Renewal Council v AEC, 5 ERC 1705; 482 F.2d 234 (2d Cir. 7/5/73).

39 Wyoming Council v Butz, 5 ERC 1844, 1846; 484 F.2d 1244 (10th Cir. 9/21/73).

40 Save our Ten Acres v Kreger, 4 ERC 1941, 1943; 472 F.2d 463; 2 ELR 20041 (5th Cir. 1/16/73); Hiram Clarke, p. 1179.

41 Hanly v Kleindienst, 4 ERC 1785, 1789; 471 F.2d 823; 2 ELR 20717 (2d Cir. 12/5/72). Cert. den. 5 ERC 1416 (5/21/73). Scientists' Institute, p. 1427.

42 Scientists' Institute, p. 1427; Arizona Public Service Company v FPC, 5 ERC 1619, 1623 (D.C.Cir. 7/30/73).

43 Hanly v Mitchell, p. 1157.

44 Hanly v Kleindienst, p. 1789.

45 Maryland Planning Commission, p. 1726.

46 Tierrasanta Community Council v Richardson, 6 ERC 1065, 1067 (S.D.Cal. 11/6/73); Upper Pecos v Stans (I), 3 ERC 1418; 452 F.2d 1233; 2 ELR 20085 (10th Cir. 12/7/71).

47 Canal Authority of Florida v Callaway, 6 ERC 1808, 1815 (M.D.Fla. 2/4/74).

48 Greene County v FPC, 3 ERC 1595, 1601; 455 F.2d 412; 2 ELR 20017 (2d Cir. 1/17/72). Cert. den. 4 ERC 1752; 409 U.S. 849 (10/10/72).

49 Hanly v Kleindienst, p. 1787; Harlem Valley v Stafford, 6 ERC 1855, 1859 (2d Cir. 6/18/74).

50 Iowa Citizens v Volpe, 6 ERC 1088, 1091 (8th Cir. 11/26/73); Citizens Environmental Council v Volpe, 5 ERC 1989, 1990; 484 F.2d 870 (10th Cir. 9/19/73; National Forest Preservation Group v Volpe, 4 ERC 1836, 1838; 352 F. Supp. 123; 3 ELR 20036 (D.Mont. 12/11/72).

51 Pizitz v Volpe, 4 ERC 1401; 467 F.2d 208; 2 ELR 20635 (5th Cir. 7/11/72). Note: Although the relevant paragraph was struck as unnecessary to the decision of this case, it indicates the Court's thinking.

52 Sierra Club v Lynn, 7 ERC 1033, 1043 (5th Cir. 10/4/74); Life of the Land v Brinegar, 5 ERC 1781, 1785 (9th Cir. 9/10/73).

53 Ely I, p. 1286.

54 Ely v Velde (II), 6 ERC 1558, 1561 (4th Cir. 5/8/74).

55 Conservation Society v Texas Highway Department, 2 ERC 1871, 1882; 446 F.2d 1013; 1 ELR 20379 (5th Cir. 8/5/71); Daly v Volpe, 4 ERC 1481, 1483; 350 F.Supp. 252; 3 ELR 20032 (W.D.Wash. 3/31/72).

56 Silva v Romney, 4 ERC 1948; 473 F.2d 287 (1st Cir. 2/2/73); Biderman v Morton, 6 ERC 1639, 1643 (2d Cir. 5/30/74).

57 Coalition for Safe Nuclear Power v AEC, 3 ERC 2016, 2017; 2 ELR 20150 (D.C.Cir. 4/7/72).

58 Silva v Romney, p. 1950.

59 Montgomery v Ellis, 5 ERC 1790, 1792 (N.D.Ala. 9/11/73).

60 EDF v Corps II, p. 1521.

61 Hanly v Mitchell, p. 1157.

62 Sierra Club v Froehlke (I), 5 ERC 1033, 1067; 359 F.Supp. 1289; 3 ELR 20248 (S.D.Tex. 2/16/73).

63 EDF v Hardin, 2 ERC 1425 (D.D.C. 4/14/71).

64 NRDC v Morton, 3 ERC 1558, 1564; 458 F.2d 827; 2 ELR 20029 (D.C. Cir. 1/13/72).

65 Calvert Cliffs', p. 1782.

66 Arlington Coalition v Volpe, 3 ERC 1995, 2002; 458 F.2d 1323; 2 ELR 20162 (4th Cir. 4/4/72). Cert. den. 4 ERC 1752, 409 U.S. 1000 (11/7/72).

67 NRDC v Morton, p. 1563.

68 NRDC v Morton, p. 1563.

69 Sierra Club v Lynn, p. 1044.

70 EDF v Froehlke, p. 1832.

71 EDF v Froehlke, p. 1833.

72 NRDC v Morton, p. 1563.

73 Silva v Lynn, 5 ERC 1654, 1657; 482 F.2d 1282 (1st Cir. 7/5/73).

74 Citizens Environmental Council, p. 1990; EDF v Corps, p. 1725.

75 Calvert Cliffs', p. 1781; Coalition for Safe Nuclear Power, p. 2017.

76 Sierra Club v Froehlke I, p. 1084.

The National Environmental Policy Act of 1969

[77] Cape Henry Bird Club v Laird, 5 ERC 1283, 1289; 359 F.Supp. 404 (W.D. Va. 4/2/73); Montgomery, p. 1792.

[78] Montgomery, p. 1798.

[79] Cape Henry, p. 1289.

[80] EDF v Corps II, p. 1519.

[81] Brooks v Volpe, 4 ERC 1492, 1498; 350 F.Supp. 269; 2 ELR 20704 (W.D. Wash. 8/4/72).

[82] Life of the Land, p. 1788.

[83] Jicarilla Apache Tribe v Morton, 4 ERC 1933, 1936; 471 F.2d 1275; 3 ELR 20045 (9th Cir. 1/2/73).

[84] EDF v Corps of Engineers (I), 4 ERC 1721, 1724; 470 F.2d 289; 2 ELR 20740 (8th Cir. 11/28/72; EDF v Corps I, p. 1922.

[85] Sierra Club v Froehlke I, p. 1067.

[86] SCRAP v U.S., 6 ERC 1305, 1312 (D.D.C. 2/19/74).

[87] Calvert Cliffs', p. 1785.

[88] Daly, p. 1485.

[89] Latham v Volpe, 3 ERC 1362, 1368; 455 F.2d 1111; 1 ELR 20602 (9th Cir. 11/15/71).

[90] Greene County, p. 1599; Harlem Valley v Stafford, 6 ERC 1855 (2d Cir. 6/18/74).

[91] Port of New York Authority v U.S., 3 ERC 1691, 1695; 451 F.2d 783; 2 ELR 20105 (2d Cir. 11/9/71); Lever Brothers v F.T.C., 2 ERC 1651, 1652; 1 ELR 20328 (1st Cir. 4/20/71).

[92] National Forest Preservation Group v Butz, 4 ERC 1863, 1865; 352 F.Supp. 123; 3 ELR 20036 (D.Mont. 12/11/72); Upper Pecos Association v Stans (II), 6 ERC 1983, 1984 (10th Cir. 7/16/74); Greene County, p. 1603.

[93] Harlem Valley, p. 1859; Latham, p. 1368.

[94] EDF v Corps II, p. 1522.

[95] Sierra Club v Froehlke I, p. 1070.

[96] Sierra Club v Froehlke I, p. 1096.

[97] Sierra Club vs Froehlke I, p. 1072.

[98] Calvert Cliffs', p. 1788.

[99] Committee for Nuclear Responsibility v Seaborg, 3 ERC 1126, 1129; 463 F.2d 783; 1 ELR 20469 (D.C.Cir. 10/5/71).

[100] National Helium v Morton, 6 ERC 1001, 1006; 486 F.2d 995 (10th Cir. 10/19/73); Monroe County, p. 1888; Sierra Club v Froehlke I, p. 1070.

[101] Sierra Club v Froehlke I, p. 1096.

[102] National Forest Preservation Group, p. 1865.

[103] Committee for Nuclear Responsibility v Seaborg, p. 1126.

[104] Hiram Clarke Civic Club, p. 1178; Portland Cement, p. 1597 (n.31).

[105] Greene County, p. 1601; EDF v TVA, p. 1857; EDF v Froehlke, p. 1830; Movement Against Destruction v Volpe, 5 ERC 1625, 1642; 361 F.Supp. 1360 (D.Md. 6/22/73); Cohen v Price Commission, 3 ERC 1548, 1552; 337 F.Supp. 1236; 2 ELR 20178 (S.D.N.Y. 1/24/72).

[106] Warm Springs Task Force v Gribble, 6 ERC 1745, 1746 (U.S. 6/17/74).

[107] Hanly v Kliendienst, p. 1793.

108 U.S. v 247.37 Acres, 3 ERC 1696, 1697; 2 ELR 20154 (S.D.Oh. 1/24/72).

109 Brooks v Volpe, p. 1498.

110 Committee for Nuclear Responsibility v Seaborg, p. 1129. See also the discussion of "Consultation with Other Agencies."

111 Hanly v. Kliendienst, p. 1793.

112 Jicarilla Apache Tribe, p. 1940.

113 Hanly v Mitchell, p. 1157.

114 Wilderness Society v Morton (II), 6 ERC 1427 (D.C.Cir. 4/4/74).

115 Silva v Lynn, p. 1655.

116 Coalition for the Environment v Volpe, 6 ERC 1872, 1880 (8th Cir. 7/31/74); EDF v TVA, p. 1853; Minnesota PIRG, p. 1701; U.S. v SCRAP, 5 ERC 1449; 93 S.Ct. 2405 (6/18/73).

117 EDF v Armstrong, 6 ERC 1068, 1070 (9th Cir. 11/9/73); Latham v Brinegar, 7 ERC 1048, 1058 (9th Cir. 8/27/74); Life of the Land, p. 1786; Minnesota PIRG, p. 1698; National Helium, p. 1005.

118 Calvert Cliffs'; Conservation Council v Froehlke, 4 ERC 2039, 2040; 473 F.2d 664 (4th Cir. 2/8/73); EDF v Corps II, p. 1523; Sierra Club v Froehlke II, 5 ERC 1920, 1924 (7th Cir. 10/2/73); EDF v Corps I, p. 1726; EDF v Froehlke I, p. 1833.

119 Conservation Society v Texas Highway Department, p. 1879; Appalachian Mountain Club v Brinegar, 7 ERC 1076, 1079 (D.N.H. 8/19/74); Sierra Club v Froehlke I, p. 1069; Conservation Society v Secretary, 5 ERC 1683, 1690; 362 F. Supp. 627 (D.Vt. 7/26/73).

120 Movement Against Destruction, p. 1640.

121 Indian Lookout Alliance v Volpe, 5 ERC 1749, 1755; 484 F.2d 11 (8th Cir. 8/22/73).

122 Boston v Volpe, 4 ERC 1337, 1339; 464 F.2d 254; 2 ELR 20501 (1st Cir. 7/17/72).

123 James River and Kanawha Canal Parks v RMA, 5 ERC 1353, 1369; 359 F.Supp. 611 (E.D.Va. 5/7/73).

124 Committee to Stop Route 7 v Volpe, 4 ERC 1329, 1334; 346 F.Supp. 731 (D.Conn. 7/7/72).

Suggested Reading

Fred R. Anderson, *NEPA and the Court,* Baltimore, Md.: Johns Hopkins Press; March 1973.

5

PUBLIC PARTICIPATION AND AGENCY COORDINATION

The NEPA process requires that Environmental Impact Statements not only be reviewed by appropriate federal, state, and local agencies but also that copies be made available for public review. The encouragement of public participation in the process is a basic tenet as well as a legal requirement of most state and federal agencies. Therefore, the EIS should be prepared with the understanding that it will be subject to public review and is to be both factual and complete.

The publication of the EIS is often preceded by citizen participation programs to foster optimum interest. These should start early in project concept and extend through the entire process with special emphasis in the planning stages. Public input should be a significant contribution in the determination of objectives and the development of alternatives.

The importance of public involvement has been recognized by both federal and state Legislatures. At the federal level Executive Order 11514 issued March 5, 1970 "places responsibility on all federal agencies to develop procedures to insure the fullest practicable provision of timely public information and understanding of federal plans and programs with environmental impact in order to obtain the views of interested parties." Most federal agencies have implemented that order by promulgating procedures to generate public participation. Most rely on the public hearing. Some, recognizing the shortcomings of a formal medium, have generated a policy of holding informal meetings in an ongoing program. Many have structured citizen advisory boards to participate in project decisions.

There are several methods of encouraging citizen involvement, but an effective program should present the issues, determine the basis for diverse opinions, and attempt to mitigate differences as early as possible.

Public faith is important; however, it must be earned. Thus, within the EIS process, efforts should be expended to generate the full spectrum of citizen opinion by fully discussing both the beneficial and adverse effects of the various alternatives, including those developed by citizen groups.

In order to generate this type of input extensive efforts have to be implemented because there is an obvious inertia on the part of most citizens against serious involvement in these procedures. Most feel that their personal activities do not allow sufficient time, that they lack expertise, or that their efforts would be wasted. This attitude tends to create a void that is filled by polarized interest groups; the "pro's" and/or the "anti's." This presents a problem because although these views are highly publicized, they often do not represent a significant percentage of the population affected. Thus efforts should be made to involve a broader cross-section of public opinion.

Several strategies have been employed to accomplish this end. These are constantly being developed, but include, at least partially:

1. Involvement of leaders from various segments and interests within the area;

2. Workshops used for disseminating information, developing at least limited expertise, and obtaining local opinions;

3. Scheduled public meetings to present an action and its alternatives;

4. Graphic project presentation with qualified attendants responding to questions.

5. Periodic newsletters describing project progress and soliciting public opinion;

6. Paid newspaper advertising to publicize public meetings and public hearings;
7. Television and radio spots as well as expanded news coverage;
8. Surveys of local officials and groups to determine local opinions;
9. Periodic questionnaires to determine public opinion and its trend;
10. Lectures describing the EIS process to various organizations so as to stimulate public interest;
11. Use of volunteers in base data collection to generate public participation and a better understanding of local environmental conditions.

Encouraging serious public involvement, however, is a relatively new process and the procedures cannot get guaranteed optimum results. Up to this point, at least, local conditions and the specific project can help determine what efforts would be most successful. Ideally, as the state of the art progresses with application and experience, procedures will be better defined and more reliable.

It is important that the availability of several copies of the EIS be advertised at least thirty days prior to the public hearing. Copies can be placed at readily accessible places such as a library for review. When citizens desire individual copies they should be provided at either a nominal charge to cover printing costs, or at no cost, depending on agency policy.

The public hearing generally required by Federal law (Chapter 14) is still the main forum for citizen input. However, it should ideally represent a summation of earlier citizen involvement so that there are no surprises. The hearing includes a discussion of engineering or logistic aspects as well as environmental issues.

In certain cases where a project is highly controversial it may be helpful to allow representative citizen groups to review a preliminary draft of the EIS before it is officially released. This will help assure that controversial issues have been adequately addressed.

It is useful to inform the public of the project time schedule. They should also be made aware that they can submit comments on the draft EIS in writing. This should be indicated on an ongoing basis throughout the public participation program.

Coordination with responsible federal, state, and local agencies is most important throughout the EIS process. Even before the assessment is begun, dialogue with responsible groups will help to determine what they are looking for in terms of environmental protection, what should be included and emphasized in the statement. Correspondence documenting their positions should be obtained and referenced and/or included in the statement.

Continuing association is crucial as these agencies can be important data sources and will be among those that are responsible for reviewing and commenting on the formal statements after submittal.[1]

The review process is extremely important as each agency reads the draft statement, addressing it primarily from its own area of expertise and responsibility. Agency comments can be meaningful as they can add data and/or ask pertinent questions that point out deficiencies in the document. In this light they have the power to question completeness or accuracy of data or conclusions, causing undue delay, which can potentially result in extended delay or even project abandonment. This power protects their respective areas of responsibility.

One note of caution: no matter how objectively an EIS is produced, in the review process the reviewer must address the document in a somewhat subjective manner, particularly in his individual area of expertise. The resulting comments may range from a difference in philosophy to simple semantics. An interview with the reviewer can generally adjust these differences to mutual satisfaction.

It should be noted that the review process is an extremely important phase and can be most productive, subject to the technical ability and philosophy of the reviewer. His familiarity with agency policy and requirements can assure meaningful responses in terms of project improvement and implementation.

Suggested Reading

Burke, Edmund M., "Citizen Participation Strategies," Volume 34, No. 5, September 1968, pp. 287-294.

Hare, Paul A., Edgar F. Borgotta and Robert F. Bales (eds.), *Small Groups, Studies in Social Interaction,* New York: Alfred Knopf; 1955.

Hunter, Floyd, *Community Power Structure,* Chapel Hill, N.C.: University of North Carolina Press; 1953.

[1] Appendix 2—CEQ Guidelines list agencies with responsibility and/or expertise to comment on EIS.

Hyman, Herbert H., "Planning With Citizens: Two Styles," *Journal of the American Institute of Planners*, Vol. 35, No. 2, March 1969, pp. 105-112.

Keene, John C. and Ann Louise Strong, "The Brandywine Plan," *Journal of the American Institute of Planners*, Vol. 36, No. 1, January 1970, pp. 50-58.

Milbrath, Lester W., *Political Participation*, Chicago: Rand McNally; 1965.

Warner, Katharine, "Public Participation in Water Resources Planning," National Technical Information Service (PB 204-245).

6

ASSESSMENT CONCEPTS

Many types of actions and almost all construction have some environmental impact, and each project has its own unique effects by virtue of geographic and environmental situations. In one case key effects might be associated with economic factors, in another, acoustic levels, and in still another, water quality; or it may be a combination of several factors. A thorough review of available engineering data, the various alternatives, and a visit to a site will establish those areas that would be subject to significant primary impacts as well as possibilities of secondary effects. Site inspection will also offer an overview for determining priorities and necessary degree of assessment.

Because of the wide variety of factors and relationships, there can be no literal "cookbook approach" to the evaluation process

As the EIS and its contents basically serve as a tool to develop an optimal action plan, it is necessary to identify significant impacts, both positive and negative. When problems are encountered the project should be reviewed to see if changes in design or location cannot be considered at this early stage, and thus mitigate negative effects at this point rather than later in the process.

Depending on the project, statement input may require the expertise of more than one individual. The breadth and range of necessary disciplines will, of course, vary with the proposed action.

One thought in relation to outside assistance in evaluation by those without engineering associations: specialists brought in for this purpose, who will have no future relationship to the project, cannot be accused of having a continued vested interest, because their responsibility ends with environmental assessment. This may not be so in the case of the proposing agency or the engineer of record.

If more than one person is required, as is usually the case, all members of the team should be supplied with all available data, maps, and as much background information as possible. When evaluators require additional information, it should be supplied as soon as possible to expedite the project. Active cooperation and liaison are extremely important in these team efforts.

The body of any assessment must include the following basic data:

1. A description of factors as they currently exist;
2. A description of the feasible alternatives;
3. The result of superimposing the project and its alternatives on these factors;
4. Suggestions to both minimize the adverse effects and where possible to even improve the environment as related to the proposal.

The Appendix contains a fictitious evaluation in the format of an Environmental Impact Statement. References will be made to this document throughout the manual. The notations indicate the appropriate section and page number in the sample. These demonstrate treatment within an EIS.

Evaluation, depending on the scope of the undertaking, should include both regional and local assessment. For instance, a highway, besides having immediate on-site impacts, could have effects over a widespread area on such factors as socio-economics, land usage, and pollution forms.

The term *downstream effects* is widely used. It relates to the results of secondary or induced effects of the project off site. A typical example is a serious change in traffic pattern as a new highspeed highway abruptly ends, and the series of staged effects that would occur in that neighborhood.

The number of factors, relationships, and interrelationships is almost infinite. Therefore, some judgment must be exercised in deciding levels of significance. This does not mean that impacts are ignored, but only that special emphasis should be placed on those that will experience the greatest change, either beneficial or adverse.

Not addressing an area that seems insignificant could result in subsequent problems, as it could well be important to others with different interests. This is an important tenet for thorough evaluation. Often when it has been ignored the results have ranged from indignation to court action. Thus, when appropriate, indicate that there would be little or no impact supported by the data to indicate that the subject had been considered.

At this point it should be pointed out that changes of greatest magnitude may not be the most significant. Relatively little change in a key factor could result in far more important effects than another of greater scope.

In determining projected changes the end effect on a feature determines the significance, and thus a relative order of importance can be developed. Sources of environmental change should be identified, which will assist in determining the potential impacts.

When assessment has established that there will be no significant impact on any of the factors associated with a project under review, according to regulations, this should be so stated. This document is referred to as a *negative declaration*. It states in essence that an environmental evaluation has determined that the proposal would have no significant effects and an Environmental Impact Statement is not required. Therefore, the action could be implemented as proposed.

The assembly of data to establish existing conditions can be expedited by using the vast store of existing data. These can be located at various local, state, and federal agencies which are generally most cooperative.

Included among these sources are:

U. S. Geological Survey
U. S. Weather Service
U. S. Department of the Interior
Department of Agriculture
Environmental Protection Agency
Local Universities

Assessment Concepts

State Environmental Agencies
Local Air Control Units
Local Planning Commissions
Bureau of Census
State Departments of Commerce and Industry

Several sections of these organizations have ongoing programs of data collection, in their own fields of expertise and responsibility, which can short-cut the time and money required to develop information for specific projects. The material has generally proven to be accurate, complete, and up to date within acceptable limits, particularly in view of the savings in time and money. These data will serve as a base to which additional information can be added, either by further research or field collection.

The proposed project and its location will establish parameters for the breadth and depth of required data. The number and condition of existing factors, type of action involved, and severity of impact all figure into this analysis. For instance, limited alteration of a small, wet weather stream does not require the same depth of assessment as impounding a major river. The number of affected features, area, and the extent of damage will generally be far less.

In situations where a multi-discipline group is used, the function of the leader or project manager becomes very important, as unit coordination in data collection, agency liaison, and writing are directly related to the ultimate value of the report.

It is interesting to note that those involved may depart from their field of expertise and editorialize on other facets in their interest and concern. Often these extraneous thoughts contain extremely valuable suggestions; however, if they stray too far from their primary area a question of subjectivity arises. This could distract from the very purpose of establishing a multi-discipline team.

When specialists go into the field for on-site investigation they should itemize each feature in the area that would be impacted by or have an effect on the project. To do this they should take into the field a map or set of aerial photographs. As they locate critical features they may annotate the map or photograph. Any specific features that are deemed important, even if they overlap, can be idenfied in this manner.

If a weighting system such as described in the Appendix[1] is

---

[1] A Mathematical Evaluation System—Appendix 6.

used, an inventory sheet will be made out for each item of interest, a sample of which is in the text of Appendix 6. This sheet will contain the classification and the specific item. It will also contain a brief description and a space for describing the impact of the proposed engineering/construction program.

The evaluation will be completed on a status quo basis and then will be reassessed to include the projected impacts. Projections should include features that will be added to the area as a result of the project. Conditions that are not amenable to specific plotting, such as the local tax base, should also be included. With this data, it is possible to compare the area in a before-and-after context and arrive at a logical conclusion as to the total effects of the proposed project.

Each evaluator should submit a written report which may or may not be appended to the EIS as reference. At this stage serious impacts will have been determined. Any significant effects should be brought to the attention of the project sponsor so that alternatives may be reevaluated and adjustments made before additional engineering efforts are expended and the EIS is formally submitted.

It is also good procedure to send a copy to each contributor to the study prior to its submission to make certain that his meanings and findings have not been misinterpreted, or that the project manager has not inadvertently omitted any meaningful data. His or her approval indicates that this facet was adequately covered.

One final note in terms of the members of the study team; they should be available for meetings and public hearings to answer technical questions as they arise. They come in this capacity neither as proponents nor opponents of the project, but only as sources of technical information. This is very important as the evaluation must be objective if it is to be meaningful. Further, the results should be thorough, addressing effects in terms of their significance. If this is the case, the primary issues will be well defined, and the scientists will only lose credibility if they assume a personal stance in relation to the project.

Almost every federal agency at this time has established content and format requirements for Environmental Impact Statements. Although the desired information is somewhat similar, emphasis, presentation, and format do vary. For instance, the Guidelines of the Department of Transportation differ from the Department of

Housing and Urban Development as their responsibilities and end goals are different. It is wise to obtain a set of guidelines from the lead agency to insure compliance. In addition, supplementary guidance, specifically designed for the prospective grantee, permittee, and other interested parties, is often available.

Included in these documents is the desired format for official submittal. Therefore, any environmental evaluation *study* should follow the outline of the specific federal or state agency, significantly reducing subsequent writing time to complete a formal statement.

Every EIS requires a description of existing environmental factors in the area under consideration [2.Pg.A-8 to 23] and the changes that will result from implementation of all proposed alternatives [Pg.A-30-38].

These options can include several approaches to the problem being addressed. These can solve the problem to varying degrees of efficiency depending on the need as well as economic and environmental costs. Included might be:

No Action

This option is interpreted as indicating that there will be nothing budgeted, constructed, or altered.

In the evaluation process, among the key issues related to this alternative are: lack of expenditures; no direct displacement of either residential or commercial/industrial firms; the lack of both temporary and permanent environmental impacts resulting from construction. In addition, however, such issues as remedial capabilities and the eventual results of this rational on both physical and socio-economic factors must also be addressed.

No-Build

This includes schemes to alleviate problem conditions in lieu of construction.

This option involves the use of solutions other than construction to relieve a problem. For instance, rather than build a new roadway, improvement of public transportation might be suggested. Strategies to discourage demand such as increased parking costs, can also be developed, as can a combination of such programs. Key issues are somewhat similar to (A) except financial requirements which could be significantly higher. The remedial effects of this option are often less direct than construction, but they can nevertheless be worthy of consideration.

Upgrading Existing Facilities

These will include a broad spectrum of construction activity, short of building an entirely new facility.

Significant effects will depend on the scope of proposed improvements. It is possible that impacts could reach the same level of importance as the construction of a new facility and should be carefully addressed. Remedial capability and costs are also to be examined in relation to both present and future environmental factors.

Construction Alternatives

These are usually interpreted as the construction of *new* facilities. Alternate types of solutions are included as well as alternate locations.

Generally the direct environmental effects of these alternatives are more obvious and involve a greater number of physical factors in particular. The following outline is typical for a major construction project and can serve as a check in compiling the data for a Statement.

    I. Description of Area
       A. Regional
       B. Local
   II. Environmental Conditions
       A. Physical Environment
       B. Socio-Economics
 III. Description of Proposed Action
 IV. Environmental Impacts
       A. Physical Environment
       B. Socio-Economics
   V. Suggestions to Minimize Negative Impacts

The depth of description for the regional and local settings will vary with the type and scope of a specific project [*1.1, Pg. A-6*]. If the proposed project tended to have regional influence on, for instance, economic factors, existing regional conditions should be explicitly described as well as those directly in the project area. On the other hand, if the impacts of the projects were purely local in nature, an in-depth discussion of regional conditions would be superfluous.

Both regional and local impacts must be addressed commensurate with their significance. *Significance* is a critical word, as the number of factors and their associations can be almost unlimited. Some line should be drawn as to the depth of assesssment.

Early in the NEPA process EIS were often considered as merely an additional "chore" to be completed and submitted as quickly as possible. Under these conditions thorough assessments were relatively rare. With the advent of more detailed guidelines and adverse judicial decisions the pendulum swung hard, resulting in many massive statements. These often contained considerable extraneous material. Such a document takes more time to develop and is difficult to review. Important considerations can become buried in the mass of detail, which can result in costly delays.

Recently the trend has been toward a more succinct document, keeping verbiage to a minimum.

In line with this philosophy determination of significant factors and effects is extremely important. Review of the project against a check list of the existing inventory and conditions should assist in this process.

Objectivity is an absolute necessity. A statement cannot be written to justify a project, but rather to clearly define positive and negative environmental impacts resulting from implementation.

Early in the system several attempts were made to use the EIS as a piece of promotional literature. Several court cases proved the futility of this ploy.

Any bias in the statement can place both the document and those who prepared it in the political arena. This reduces credibility and often generates unnecessary opposition. Preparers, reviewers, and the public have all become more knowledgeable and sophisticated.

The evaluation should include *positive* as well as *negative* effects so that the assessment includes a complete picture of the proposed project.

Secondary effects are extensions of primary impacts. For instance, the volume capacity of a floodplain can be seriously reduced by construction within the area. When this occurs, land adjacent to the river upstream would be more vulnerable to flooding as the water cannot get through the partially blocked segment at its former rate. Thus, although located some distance from the project, this area could suffer serious loss of life and property. In this vein such impacts are often called "upstream" or "downstream" effects. These are also termed "domino" effects as they are generated by other impacts.

A lack of concern with second level impacts has often been challenged. The extended view is important and those impacts should be considered as related to the project. This rationale is proving to be an important phase of impact evaluation as it has been discovered that secondary effects can sometimes prove to be far more critical than primary effects.

Secondary effects are generally less obvious and often extend over a much broader area. They should be established by following primary impacts through to a logical conclusion.

Information such as long lists of biota are sometimes necessary; however, they serve no purpose if they merely add bulk to the assessment. Explicit detail should be presented only when it is pertinent [2.6 & 2.7, A.19-20, A.20-23].

The level of vocabulary is important. It will be reviewed by a wide range of readers, from competent professionals to untrained but interested citizens. Although the document is basically a technical report it should be presented in a readable style and couched in language that is readily understandable. Technical reports as data sources can be appended to the report to support its rationale and conclusions.

Tables and figures should be prepared in a clear and concise manner so that factors and alternatives can be easily compared [A.3,4]. Graphics are important as they can clearly demonstrate a wide range of data and serve as reference to the report.

Considerable efforts have been devoted to developing meaningful systems, so that mathematical comparisons of alternatives can be established.

Weighting approaches such as McHarg's system, check-off systems (United States Geological Survey), and several computer solutions have been painstakingly developed and each has been an important attempt at making the assessment process more stable. However, in my opinion, the concept contains some very real pitfalls, because:

unless the identical weighting system is used for every project of the same type, the listing of factors to be weighted can serve to control final results;

not only are there a large number of factors to be considered, there are also the interrelationships. This could result in an unwieldy number of comparisons.

The same weighting system must be used for the same type of project or an "apples and oranges" type of assessment could occur. Although some factors can be mathematically determined, such as cost, user benefit ratios, comparative displacement, and acreage demands, there are many factors that cannot be established without subjective input. As viewpoints vary, it is apparent that it would be most difficult to develop universal standards. The use of such a system should be approached with caution.

If a simple weighting system is desired, one might simply use large pluses for major beneficial impacts, large minuses for critical negative impacts, zero for no impacts and small pluses or minuses to represent less significant effects. A simple totaling of these will give at least a rough comparison of the environmental effects without including any numerical figures.

If a more complicated system is required, one approach to the system is presented in the Appendix.[1]

There is one additional note in relation to the evaluation process. Although the procedure has yet to be widely adopted it would seem meaningful to include in the planning process some wherewithal to maintain an ongoing surveillance of estimated effects during and even after construction to:

1. Determine that acceptable procedures for environmental protection were observed;
2. Determine that impacts occurred as projected;
3. Observe the resulting eco-system to determine the effectiveness of protective measures;
4. Recommend additional procedures to improve design and construction procedures in similar projects.

With the large variety of conditions and factors to evaluate, various assessment methods have been developed by practitioners in public agencies, universities, and private practice. These were

[1] Mathematical Evaluation System—Appendix 6.

the result of expertise, experience, and specific interests. Thus, other approaches are available which could be more compatible to the reader's philosophy and situation. Most are available and will ably serve those faced with the responsibility of completing an evaluation for a specific project under the administration of a specific agency. Notable examples include the system and matrix created by the United States Geological Survey and the Soil Conservation Service.

Suggested Reading

The Council of Planning Librarians recently published an annotated bibilography (Exchange Bibliography #691). It is entitled "Environmental Impact Assessment Methodologies" and was compiled by Richard C. Viohl, Jr. and Kenneth G. M. Mason of the National Park Service, U. S. Department of the Interior.

The bibliography is the most complete listing available and offers a variety of approaches to environmental assessment. It offers a synopsis of each listing. The following is an example of the format.

PART A: METHODOLOGIES AND SYSTEMS FOR
ENVIRONMENTAL IMPACT ASSESSMENT

"An Approach to Assessing Environmental Impact," David W. Fischer and Gordon S. Davies, *Journal of Environmental Management,* Ontario, Canada: University of Waterloo, Vol. 1, No. 3, July 1973, pp. 207-227, 8 figures.

A method of environmental analysis is proposed that expands the concept of environmental matrices. Three steps are used to identify and evaluate environmental feasibility: (1) a matrix for environmental baseline evaluation in terms of importance, present condition, and management, (2) an environmental compatibility matrix in terms of introduced activities, and (3) a decision matrix for evaluating alternatives available (no project, structural and non-structural alternatives, locational alternatives). This methodology relies heavily on the use of multidisciplinary teams to provide value scales and to perform the environmental analysis.

7

POLLUTION POSSIBILITIES

Various types of factors may be addressed in environmental impact evaluation depending on the proposed action.

Some planning efforts and legislative actions are capable of having as far-reaching effects as construction. However, because construction is more dramatic in its effects and because major projects are subject to more intense review, discussion in this manual will primarily relate to impacts associated with new facilities.

Among physical effects are the possibilities of increasing pollution levels. Others include:

Loss of natural resources;
Loss or change of habitats and changes in the ecological web;
Loss or development of open space and scenic areas;
Changes in land use, which will be addressed later in the text as will socio-economic effects.

Pollution potential has been treated separately in this manual because the effects are more spectacular, and more progress has been made in their identification and quantification. Standards are better defined and highly-developed pollution abatement programs have been underway at almost every level of government.

The following overview can serve to bring into perspective potential effects of a project.

1. The activities associated with a project will help determine possible impacts. These could include:

Demolition	Channelization
Renovation	Sewerage
New construction	Floodplain alteration
New/additional equipment	Watershed alteration
Additional vehicular activity	Industrial discharge
Water impoundment	Landfill
Mining	

2. Negative impacts on the following factors could be interpreted as increasing pollution activity. Both *temporary* effects, i.e., those occurring during the construction period, and *permanent* effects should be examined and addressed.

Acoustics
  Noise level increases

Air Quality
  Chemical additions
  Odors
  Dust
  Smoke

Surface water quality
  Sedimentation
  Chemical effluents
  Thermal alterations
  Sewage effluents

Ground Water
  Water table level changes
  Leeching
  Other chemical (Bio-chemical) additions
  Loss of natural water absorption area

Land Destruction
  Acreage involved
  Intensity of destruction
  Reclamation possibilities

A brief description of these possibilities and some basic information follows.

ACOUSTICS

Sound is the result of mechanical vibrations in mediums including gases, fluids, and solids, which are characterized primarily by frequency, amplitude, and to a lesser degree, phase.

The level of sound is measured using a relative scale of sound pressure entitled the decibel (dB) scale. Measurements are generally made on the A scale; thus readings are usually presented in dBA's.

As a form of environmental pollution, noise is of increasing concern to physical and social scientists [2.3, A-9]. It has been shown that undue decibel levels create stress, emotional disturbance, reduction in operating efficiency, and loss of property values [4.3, A-32]. Therefore, standards are being established for maximum decibel levels by several federal and local agencies (Tables 1 and 2).

The L_{10}, L_{50}, and L_{90} referred to in FHWA's Policy and Procedure Manual 90-2 are statistical levels which represent those levels which are exceeded 10, 50, and 90 percent of the time respectively. Note that the specified maximum levels vary with land use. This is practical as it is not as restrictive as across-the-board requirements that have no logical rationale.

In order to determine significant acoustic changes either during construction or as a result of a completed project, existing levels should be monitored by sound level meters. Depending on the type of project and agency requirements, the length, frequency, and number of receptor stations will vary. Levels should be measured so that ambient levels are established as well as levels at key locations potentially subject to significant increases. The times of day will depend on the locale and its activities; however, the high, low, and critical levels should be included. For example, although rush

FHWA ACOUSTICS SPECIFICATIONS

Table 1. Design Noise Level/Land Use Relationships

Land Use Category	Design Noise Level—L_{10}	Description of Land Use Category
A	60 dBA (Exterior)	Tracts of lands in which serenity and quiet are of extraordinary significance and serve an important public need, and where the preservation of those qualities is essential if the area is to continue to serve its intended purpose. Such areas could include amphitheaters, particular parks or portions of parks, or open spaces that are dedicated or recognized by appropriate local officials for activities requiring special qualities of serenity and quiet.
B	70 dBA (Exterior)	Residences, motels, hotels, public meeting rooms, schools, churches, libraries, hospitals, picnic areas, recreation areas, playgrounds, active sports areas, and parks.
C	75 dBA (Exterior)	Developed lands, properties or activities not included in categories A and B above.
D	—	For requirements on undeveloped lands see paragraphs 5.a(5) and (6) of PPM 90-2.
E *	55 dBA (Exterior)	Residences, motels, hotels, public meeting rooms, schools, churches, libraries, hospitals, and auditoriums.

hours (7:00–9:00 A.M. and 4:00–7:00 P.M.) are generally the period of higher levels on a suburban roadway, and 2:00 A.M. would probably represent the lowest, the critical period for an adjoining school would occur during school hours.

Length of measurement will also vary, but ten continuous minutes should be considered a minimum.

Superimposing the acoustic implications of the proposed action on existing levels allow for projections as to future levels both during construction and after completion. First, the number of pieces of equipment, traffic or other noise sources should be determined. Then, library or manufacturers' data should be consulted as to the decibels generated and these added to current figures. If this data is not available, direct measurements can be made. It should be mentioned that the addition is *logarithmic* and not simple. The

Table 2. External Noise Exposure Standards for New Construction Sites

General External Exposures dBA	Airport Environs	
	CNR Zone *	NEF Zone *
UNACCEPTABLE Exceeds 80dBA 60 minutes per 24 hours Exceeds 75 dBA 8 hours per 24 hours (Exceptions are strongly discouraged and require a 102(2)C environmental statement and the Secretary's approval)	3	C
DISCRETIONARY—NORMALLY UNACCEPTABLE Exceeds 65 dBA 8 hours per 24 hours Loud repetitive sounds on site (Approvals require noise attenuation measures, the Regional Administrator's concurrence and a 102(2)C environmental statement)	2	B
DISCRETIONARY—NORMALLY ACCEPTABLE Does not exceed 65 dBA more than 8 hours per 24 hours		
ACCEPTABLE Does not exceed 45 dBA more than 30 minutes per 24 hours	1	A

(Measurements and projections of noise exposures are to be made at appropriate heights above site boundaries)
* See Appendix 2 for explanations of Composite Noise Rating (CNR) and Noise Exposure Forecast (NEF).

Highway Research Board has published a manual for projecting highway associated levels.[1]

The following two considerations are important: the increase in ambient levels, and more important, the localized impacts. Not only the difference in levels but also the immediate locale are important. An increase within an area with high existing levels will have less effect than that same increase in a more quiet zone where the difference would be much more noticeable. Furthermore, in an area that is relatively serene by comparison even a smaller increase would result in far more significant effects.

There are various methods for attenuating acoustical impacts. These include:

[1] "Highway Noise," (NCHRP #117), National Cooperative Highway Research Program.

A. Dense planting (100 feet of dense vegetation reduce levels up to approximately 5 dBA's)
B. Various forms of earth or fence-like barriers;
C. Constructing the facility either below or above grade;
D. Moving the noise source a greater distance from a potentially impacted receptor as there is a 3 dBA reduction in each doubling of distance;
E. Use of noise-quieted equipment by contractual arrangements to reduce levels during the construction period.

AIR QUALITY

Pollution levels in ambient air and their effects on the environment have become a major concern of agencies at every level of government [2.2, A-9]. Of particular importance is air quality in the metropolitan areas that are subject to excessive industrial and vehicular emissions [4.2, A-30,31]. The results of heavy pollutant loads are obvious, particularly in larger cities where smog, darkened skies, and unpleasant odors are common occurrences. Conditions are serious enough in these locales to motivate the Environmental Protection Agency and related state agencies to create priority areas requiring controls to alleviate pollution levels.

To improve what were steadily worsening conditions the federal legislature passed the Clean Air Act of 1970 [2] which empowered the Environmental Protection Agency to become the leading enforcement agency dealing with air quality, to require that states establish implementation programs to reduce pollution levels, and to make EPA responsible for assisting them in those efforts.

State and local government agencies have been simultaneously created which also have powers of regulation and enforcement as well as research capabilities.

Federal Air Standards particularly from vehicular sources (Table 3) have been established which have served as a basis for a series of actions and strategies aimed at reducing emissions from existing facilities; and at limiting new sources of pollution.

Sophisticated monitoring networks have been generated to continuously record pollution levels and enter those data into retrieval systems.

[2] Clean Air Amendments of 1970 (P.L. 91-604) December 31, 1970.

Pollution Possibilities

Double Air Inversion Over Los Angeles, Caliofrnia, 1972
(Courtesy: EPA–Documerica, Gene Daniels)

Table 3. FEDERAL AIR QUALITY STANDARDS

Federal air quality standards for total oxident, oxides of nitrogen, carbon monoxide, and hydrocarbons. (Primary and secondary standards are similar for these pollutants).

Pollutant	Federal Air Quality Standard
Total Oxidant	0.08 ppm for 1 hour not to be exceeded more than once per year.
Oxides of Nitrogen	0.05 ppm annual average
Carbon Monoxide	9.0 ppm for 8 hours
	35.0 ppm for 1 hour
Hydrocarbons (non-methane)	0.24 ppm for a 3 hour period from 6 A.M. to 9 A.M.

ppm = parts per million

Several court decisions have strengthened the powers of responsible agencies by key decisions and firmly established agency responsibilities in enforcement of emission abatement measures. These have put teeth into programs that are and will involve staggering sums of money by industry in order to insure compliance. In some cases it has and/or will prove to be more feasible to close down air-offending facilities. This obviously has a serious economic and social impact on the whole community. In all fairness both public agencies and industry have mediated in an attempt to arrive at practical solutions. These joint efforts must succeed to insure correction without economic upheaval.

There are several types of pollutants generated by different sources. Industrial emissions, which are generated by pinpoint (stack) sources include such materials as:

Particulates	Lead
Beryllium	Fluorides
Sulphates	Sulphur dioxide
Sulphuric Acid	

Not only do excessive amounts of such chemicals have effects on public health but they lead to other effects. For instance, sulphur emissions in association with water creates sulphuric acid of varying strength which corrodes and dissolves many types of materials including, among others, metals, natural stone, and concrete.

Construction activities, soil removal, quarrying, and similar activities are also sources of pollutants; so is fire, whether in a boiler or in the field.

Regulations to prevent or reduce the level of these emissions have been established on local and state, as well as federal levels. When contemplating any project that might tend to develop or increase a pollution, regulations should be reviewed to insure the ability of the proposed project to comply with standards.

Point sources can be controlled by several types of abatement controls available on the market which can be engineered for special adaptation.

Another major source of air pollution is vehicular emission. As the number of both automobiles and trucks has increased, the load has also been increasing.

Federal legislation has set time limits for the manufacturers to reduce vehicular emissions by 1975;[3] however, although several devices have already been installed, it has become apparent that the 90% reduction goal cannot be reached within the allotted time and already the Environmental Protection Agency has agreed to one extension. It should be pointed out that extenuating conditions such as the energy crunch forced re-evaluation of previous rationale in all forms of energy generation which have not at this point been completed.

Emissions generated by vehicles include carbon monoxide, non-

[3] Clean Air Act, 1970.

methane hydrocarbons, oxidants, and nitrous oxides, as well as other photo-chemical pollutants.

The volume of traffic, truck-auto mix, mode of driving (speed, stop, and go), meteorological conditions, and physical situation of the roadway (sections in cut, fill, and elevated portions) all have a direct effect on the level of emissions.

In areas that have been designated as high priority, or subject to excessive levels, determination of existing levels is accomplished by actual field monitoring, which establishes current parts per million (ppm) of the various pollutants.

Computer models have been developed that will take into consideration such factors as monitoring results, change in traffic volume, driving modes, age of vehicles, and the quality of their abatement equipment. This offers the ability to make reasonably accurate projections for future conditions if a proposed highway is completed or not completed. Typical programs include the California Line Source Model and EPA's HIWAY Model.

For those areas where conditions are not so severe as to warrant such in-depth data, the aforementioned programs are used with assumed existing pollution loads that will accomplish the same end except in degree of accuracy. These are more theoretical and do not produce as sensitive a level of results. However, they take less time, equipment, and cash outlay, and are adequate in less sensitive situations.

Several strategies have been developed to alleviate the volume of traffic and the ensuing emissions apart from mechanical improvements. These include greater proportion and dependence on mass transit in every form: the promotion of efficient car pool programs; staggered work shifts; and increased cost and less availability of all-day parking; and finally even the use of bicycles. Several of these have reached various localized levels of success; however, all require a tremendous amount of promotional and educational effort to overcome inertia and the traditional American dependence on one's own personal mode of transportation.

The solutions to the many problems related to air quality are not yet complete. However, a great deal of effort is being expended that will eventually result in the determination of realistic emission level standards. Consideration is underway to establish practical controls over indirect sources as, for instance, supermarket parking lots; however, other related problems, such as economic conditions, are delaying final ordinances. In the meantime, it is important to address the effect of a proposed project on potential air quality during the assessment and engineering phases.

WATER QUALITY

The quality of our surface water has been a subject of increasing serious concern [2.4, A.11-18]. Thus, federal, state, and local governments have passed legislation, organized regulatory agencies, and established basic standards [4.4, A.32-34]. These programs have been devised to prevent additional pollution and also to improve existing quality when it has fallen below standards.

There are several bills directly related to this topic, and the

Discharge From Sewage Treatment Plant In the South Platte River
(Courtesy: EPA–Documerica, Bruch McAllister)

Federal Water Pollution Control Amendments of 1972 is as important a piece of legislation as any.

Several federal and state agencies have established basic standards and the responsibility for their enforcement. On any project it is wise to call the responsible agency to determine what requirements must be met in relation to a proposed project.

Although there are many factors determining water quality the following parameters are most often used.

> pH: This is a comparative measurement of the acidity-alkalinity relationship. Normal accepted ranges will vary between 6.0 and 8.5.
>
> Total Alkalinity: This generally serves as an indicator of aquatic productivity.
>
> Dissolved Oxygen: This also serves to determine the ability of the water to support aquatic life. Levels in the magnitude of 4 to 5 mg/l [4] are considered satisfactory whereas near saturation at levels of 8 to 9 are considered excellent.
>
> Chlorides: Their presence is related to such factors as highway runoff where the material is used in snow/ice control, and in the treatment of sewage.
>
> Temperature: This is an extremely important indicator of water quality, particularly if viewed in conjunction with dissolved oxygen levels. Any serious variations from the local norm can seriously affect resident species and the ecological system.
>
> Biochemical Oxygen Demand (BOD): This factor represents the organic waste load. Ranges of from 1 to 2 mg are generally considered satisfactory.
>
> Chemical Oxygen Demand (COD): This represents the chemical organic load and the oxygen demand of that material. In essence this figure represents an extension of and is studied with the BOD level.
>
> Coliform Level: This represents the amount of bacteria generated from both the waste of warm-blooded animals and from the soil. Fecal coliform counts serve as an indicator of the level of sewage pollution contributions as they derive primarily from human waste.

---

[4] Milligrams per liter.

Although state standards vary, no more than 1000 to 2000 counts per 100 ml [5] are allowed in water that is to be used for swimming activities.

Nitrogens and Phosphates: These serve as indicators of eutrophication [6] which is an over-supply of nutrients with a corresponding depletion of oxygen, particularly in the lower levels.

Suspended Solids: This is a measure of particles maintained in suspension in the body of water. The greater the number, the greater the turbidity.

Total Dissolved Solids: This level serves as an indicator of inorganic pollution of which brine might serve as an example.

Traces of heavy metals can also be measured. Even small amounts of some of these could be significant.

It should be pointed out that the presence of most of the materials and/or conditions just described need not be serious. It is only when the levels exceed certain limitations that they are to be classified as pollutants. In addition, not all are man-produced. For instance, it is known that some species of ducks defecate approximately twelve ounces each day. If a large number are in residence near a small lake, it becomes obvious that nutrient levels will be excessive.

In terms of human influence on water quality, several of our activities are major contributors to high pollution levels.

For instance, industrial wastes have long been recognized as significant sources of chemical, both organic and inorganic, substances. Existing controls are now tailored to limit the discharge of injurious material.

Agricultural activities have been proved responsible for an over-abundance of nutrients as well as various pesticides, herbicides, and fungicides.

Sewage treatment facilities have often not been adequate but

[5] Total coliform count.

[6] Eutrophication commonly associated with human activities is actually a normal occurrence related to the aging process of a body of water. It is complex and with no unnatural influence generally presents no problems. However, when man's activities intrude, the process which naturally takes centuries to occur, can be critically foreshortened.

are currently being upgraded on a rather large scale with federal funding.

Certainly mining activities have contributed to the high acid content of many streams as the result of the interaction of sulphur and water.

Power plants utilizing water on a once-through cooling cycle discharge a higher temperature effluent than the receiving waters, causing a definite effect on local aquatic life.

Highway deicing methods have proven to offer measurable amounts of contaminants.

Construction activity yields high concentrations of sediments where sound erosion control precautions have not been implemented.

These will serve as examples, but other sources related to specific projects can be identified.

The sources can be either *point*, such as a discharge pipe, or *non-point*, such as agricultural runoff. It is obvious that the quantity and concentration of point sources are easier to determine. However, programs are being developed to quantify other more complicated sources.

GROUNDWATER

Effects on groundwater are extremely important [*2.4, A-11-18*]. This is water that is present beneath the surface of the earth. It is confined by an impervious layer of rock and its highest point is controlled by the volume of existing water which seeks its own level. The high level is referred to as the water table [*4.4, A-32-34*].

An estimate in 1954 indicated that groundwater use in the United States was approximately forty-one billion gallons per day or one sixth of the total water utilized in this country. No reduction in usage is contemplated which reinforces the claim to the importance of groundwater and the significance of the water table, which is, to repeat, the upper limit of the saturated zone.

Logically the aquifer, as the stratum that contains and carries the water, should be safeguarded so that it will continue to function in those capacities and insure protection for the water against pollution from outside sources. Damage possibilities include cutting into the strata, which would allow a loss of volume and/or the intrusion of pollutants.

Within this reasoning special care must be taken with such projects as sanitary landfills, strip operations, changes or movements of septic tank systems, tunneling and dredging; in other words, any activity that would tend to disturb the water-carrying strata and subject that water to a reduction in quantity or quality.

LAND DESTRUCTION

Certain actions are prone to destroying land, particularly when no reclamation steps have been included [2.4, A-15, 4.4, A-33]. Such projects include, among others less spectacular:

1. Quarrying and strip mining;

Aerial Photograph of Strip Coal Mining Operation
(Courtesy: EPA–Documerica, Bill Gillette)

2. Undue excavation of areas for fill, slated for abandonment and thus loss of even marginal use once the process is complete;
3. Wanton destruction of prime agricultural acreage or prime timber forest areas;
4. Extensive concrete cover for marginal purposes, which also destroys land by making it economically unfeasible to reclaim.

Those activities related to excavation should be restricted unless proper reclamation procedures to insure a return to productiveness are incorporated into project planning. Federal and many state governments are proposing and, it is hoped, passing stringent legislation to insure this type of protection. However, those Acts will require time for enactment and many of the requirements can be made less stringent because of economic impacts, the shortage of energy sources, and successful lobbying tactics.

Nevertheless, the long range effects of land loss are significant and should be closely assessed in evaluation of projects that indicate the possibility of generating significant loss of land.

Mitigation suggestions should include a plan and schedule for reclamation. There is a possibility that at some time in the future the posting of a bond to insure these conditions will be required; however, at present at least the destructive nature of the project should be directly addressed.

It is interesting to note that with imaginative planning, the land owners would be able to remove the material they are seeking, but also when they are finished, own a piece of property that is an economic asset, capable of many productive uses.

Suggested Reading

Acoustics
   "Handbook of Noise Measurement" 7th Ed.
   General Radio Corporation
   Acoustic Noise Measurements, 1967
   Bruel and Kjaer Company.

Air Quality
   Arthur C. Stern, ed., *Air Pollution,*
   Volumes 1, 2, & 3, New York:
   Academic Press, 1968.

   Environmental Protection Agency and Federal Highway Administration Regulations and Publications.

Water Quality
   P. H. McGauhey, *Engineering Management of Water Quality,* New York: McGraw-Hill, 1968.

   U.S. Water Resources Council, "Principles and Standards for Planning Water and Related Land Resources," *Federal Register* Vol. 38, No. 174, Part III, Sept. 10, 1973, pp. 4-9, 13-14, 19, 26, 33-34, 59-75, 91, 98-99, 108, 113-115 (of orig. text).

U.S. Senate, "Policies, Standards, & Procedures in Formulation, Evaluation, & Review of Plans for Use & Development of Water & Related Land Resources," Senate Document #97, 87th Congress, 2nd Session, May 29, 1962.

A *Methodology for Assessing Environmental Impact of Water Resources Development*. Final Report for Dept. of the Interior by Environmental Impact Center, Inc., Cambridge, Mass., Nov. 1973, 148 pp., 32 figures, 19 tables.

Groundwater

Edward E. Johnson, Inc., "Groundwater and Wells," St. Paul, Minnesota, 1966.

David Keith Todd, *Groundwater Hydrology,* New York: John Wiley & Sons, 1963.

8

ECOLOGICAL AND ENVIRONMENTAL INVENTORY

Ecology, by definition, is the relationship of an organism to its environment. An ecological system is the sum of existing interrelationships in that system. For instance, the entire globe is an ecological system. On the other hand, a subsystem exists within a watershed or a bog. Regardless of its size, each unit contains an inventory of items that interrelate, each contributing to the total. Current common usage of the word "ecology" primarily refers to man's relationship with his environment. Scientists, however, are aware that an entire system has to be protected in order to maintain the balance. Thus, the inventory has to be evaluated in order to determine potential losses or alterations that would result in either negative or positive effects.

The following classification check-off outline lists basic factors that could be related to an action or project. Each one of these items would have a different value and importance depending on quality, quantity, and relationships with the other factors.

There are no blacks or whites but rather shades of grey in this type of evaluation, and relative values should be assessed. It should be noted that one purpose of evaluation is to mitigate negative and/or increase positive effects of each alternative to aid in selection of the optimum course of action. Thus no hard or fast rules can be developed. The environmental inventory consists of the features that exist within an area. In establishing such an inventory, both quantitative and qualitative facets might be included. The following outline presents many of the numerous types of items that might be found in environmental inventory.

Topographic and Scenic Items
    Basic Geology
    Prominent Elevations
    Caves and Sinkholes
    Cliffs and Outcroppings
    Stone and Mineral Collection Areas
    Ore Outcroppings
    Agricultural Projects
    Soil Conservation Projects

Water Resources
    Drainage Patterns
    Significant Streams
    Ground Water
    White Water
    Lakes and Ponds
    Beaches
    Wetlands
    Dams
    Reservoirs
    Mill Ponds
    Canals and Locks

Vegetation Resources
    Virgin Timber Stands
    Rare Timber Remnants
    Reforestation Projects
    State Forests
    National Forests
    Orchards
    Nurseries
    Interesting or Rare Plant Areas

Game Inhabited Areas
Unique Habitats

Historical and Cultural Resources
Covered Bridges
Old Mills
Trading Posts and Taverns
Old Mining Facilities
Historical Homes and Buildings
Old Forts
Museums
Battlefields
Old School Buildings
Old Churches
Cemeteries
Libraries

Archeological Resources
Burial Grounds
Dig Sites

Wildlife Resources
Big Game
Small Game
Water Fowl
Hatcheries
Game Fish
Game Fowl

Supporting Biological Web
Terrestrial—
  Vegetation
  Fauna
Aquatic—
  Flora
  Fauna

Recreational Facilities
Hunting Areas
Water Orientated

Natural Resources
  Mineral
  Agricultural
  Timber
  Soils
  Water Source

Because this inventory, by sheer number of interrelationships, becomes somewhat confusing, it is worthwhile to present a brief rationale as to their importance and some of their possible relationships.

Topographic and Scenic Items

The *basic geology* of an area, both surface and subsurface, can be an important consideration in construction, but it may also be of aesthetic or instructional interest when exposed [2.5, A-18; 4.5, A-34]. Often road cuts are raw from an aesthetic point of view; however, such man-made features can be exceedingly interesting by exposing the underlying geology of a region which is otherwise buried by a mantle of soil and vegetation.

Outcroppings, on the other hand, may have great aesthetic and scenic value, often preserving unusual vegetation types and species not found elsewhere.

Prominent elevations often command an imposing view of a local or regional area and could play an important role in the local micro-climate by eliciting updrafts that result in summer convectional storms.

Caves and cliffs have an inherent value because of man's fascination with hidden and dramatic features of the landscape. However, it should be pointed out that these values have a wide range. For instance, in a limestone area that is rife with caves of various sizes, shapes, and contents, a specific cave would have to be particularly spectacular to be considered valuable. But, if the cave were the only one in a large area, it would be significant by virtue of its very existence. Even man-made caves, such as abandoned mine shafts or tunnels may have developed into important habitats for bats and other species partial to that type of environment, which due to their insectivorous feeding habits, could be quite useful from biological control standpoint in local ecosystems.

Ecological and Environmental Inventory 67

Large agricultural areas and land conservation projects are important both from the standpoint of physical attractiveness and their economic value to the area. The increasing use of farm land, obviously, falls victim to construction projects of all types because it has been available at lower acquisition cost than most other types of land, while still allowing for a reasonable profit to the owner. Productive farm land is quite finite in extent and great care should be taken in the siting of projects to utilize less productive land. Once taken, farm land cannot be restored to crop production. At the present, our margin of food production over consumer needs is quite large; but it will not always remain so.

Water Resources

Since water in any form is a basic requirement for all life, it requires special consideration and protection [2.4, A-11-18; 4.4, A-32-34]. Because this is a pressing matter, special agencies have been established for water quality protection. However, in the present context, we are primarily concerned with specific aquatic features, rather than water quality itself.

Natural drainage patterns are a basic consideration in every engineering study, but have to be considered from another standpoint in an ecological study. Since the drainage system is part of the matrix for local ecological systems, any alteration significantly changes that system. Wildlife losses must be projected and the limits of the new system and its capability of sustaining resident flora and fauna must be determined.

Channel changes have been relatively common, usually involving the straightening of a water course. These often have resulted in significant ecological impacts [3.1, A-24; 4.4, A-33-34; 4.7, A-35]. Meandering streams have evolved over long periods of time and serve not only vital ecosystem functions but in flood protection as well. It is important to realize that flood plains serve a most important function, and intrusion or destruction could have widespread, long-range effects.

White water is the term used for rapids and other turbulent areas of streams. They may be common or rare, depending on the stream gradient, but are almost always physically attractive. In addition to their aesthetic contribution they have potential recreational use by canoeists, depending on water speed and depth. Although a hindrance to navigation by power boats, they are recreation attractions to users of canoes, rafts, or floats.

Natural impoundments, such as lakes and ponds, are generally worth protecting, although their value is usually directly related to their specific settings and environmental relationships.

Many man-made features such as dams, reservoirs, canals, and locks, constructed as the result of public need, often have become well integrated with the environment and have attained the status of natural features. As such, they could be considered as natural features rather than expendable man-made objects.

Wetlands such as bogs, swamps, and marshes, have finally become recognized as prime environmental features. Several states have enacted legislation for their protection. However, all such areas are not of equal ecological value; much depends on their origin, quality, use, and abundance. Research is being conducted to redevelop by marsh restoration programs areas that have been rendered valueless by pollution and land use changes.

Vegetation Resources

The value of vegetation in the environment is undeniable because of its role as a supplier of fuel and pulp products as well as the shelter it provides for a prodigious number of organisms [2.6, A-19,20; 4.6, A-34]. In addition, its aesthetic importance is obvious. However, the type and maturity of the vegetation varies broadly; for instance, an old field of weeds versus a virgin beech-maple stand. One basic factor in determining the value of the vegetation is replacement time. For instance, a stand of virgin timber of considerable economic and aesthetic value may require hundreds of years to replace; a stand of pulpwood, twenty years. Other vegetation types, particularly those representing the early stages of succession, may be either expendable or easily replaced.

Naturally, interesting and rare plant communities should be given greater priority than more common types that can be regenerated with ease—for example, a stand of hemlock in an oak-hickory forest, a patch of natural grassland in a forested area, or a marsh in an otherwise well-drained site.

National and state forests, as well as game lands and parks, superficially offer little to the economy because they are publicly owned. Thus, in the past it was reasoned that they would keep acquisition costs to a minimum, but they have been infringed upon by many types of construction. Section 4(f) requirements and a recent Supreme Court decision, however, have made this practice much more difficult. Still, wilderness areas are becoming more scarce and, hence, more valuable to society. Intelligent use of these has become a necessity. Encroachment should be allowed only when in the national interest, and then approached with caution, care, and conscience.

Reforestation projects, nurseries, and orchards, which generally represent meaningful and financially responsible land usage, should not be routinely destroyed simply because they can be acquired more cheaply than land bearing buildings. If taken, replacement cost should include and represent total ecological value.

Historical and Archeological Resources

Discussed in chapter eleven

Wildlife Resources

The wildlife population of the United States is in a state of precarious equilibrium with the landscape that supports it. As land is taken, species may either disappear or be forced into smaller or more remote areas. Ultimately, if the population of man continues to increase, many forms of wildlife will be literally eliminated to make room for domesticated plants and animals, and finally, man himself. There is in fact a list of rare and endangered species.[1] In its effort to protect those species the government prohibits any action including the removal of advantageous habitat, that would tend to endanger any of their number.

Most species require natural land in varying quantities. Most construction not only reduces habitat acreage but also fosters the presence of man, the predator. Not only land types are subject to these effects, but birds and aquatic life as well. Fish populations, for example, are seriously affected by channel changes, construction-generated pollution, dams and point and non-point i.e., agricultural runoff, discharges of pollution.

The federal and many state governments have agencies estab-

---

[1] U.S. Dept. of Interior, Bureau Sport Fisheries & Wildlife, "Rare and Endangered Fish and Wildlife of the U.S." Resource Pub. 34 (Revised 1972) Washington, D.C.

lished for the management and protection of wildlife; however, these agencies are limited in their jurisdiction when dealing with general construction by private organizations.

If there is to be an effect on the local wildlife population, it is important to consider if and where those species will migrate. Alternate habitats should be relatively close and capable of offering the necessary food and shelter. If it is possible, the establishment of such a managed area as part of the construction project may be a reasonable trade-off.

Supporting Biological Web

Although both flora and fauna have been previously listed and discussed, the sections were devoted to the more dominant or spectacular species, but their supporting web is certainly as critical, because with its loss, the more obvious types would be reduced in number or even disappear [2.7, A-21].

These species include the food and base habitat requirements. In the aquatic environment benthic organisms as well as flora and smaller fish species act as support for fish life. In addition, their presence with other species, such as algae, contribute to water quality control.

Insects, small plants, and even small mammals serve as food sources for larger animals as well as assist in controlling other species by their feeding habits. Vegetation is also controlled in a natural manner.

These interrelationships are quite complicated; however, once the species and numbers have attained an equilibrium in a given area, a stage of what is termed ecological balance has been established. This relationship can apply to a specific bog or the entire globe, and it is any significant change in the balances that scientists are so concerned about.

The relationships are never static, and they can withstand minor intrusions. However, a severe disruption can create unfavorable conditions for certain species. Those either migrate or perish, and other inhabitants fill the void to create yet another system. Recovery and development are deliberate, generally depending on the force of the impacts.

It should be noted that systems can also be upgraded with management. Nevertheless, when possible, it is obviously best not to disturb a balanced system and changes should be examined in terms of ecological consequences.

It should also be noted, and with great emphasis, that this section is grossly over-simplified because this is a manual and not a biological text. The more valuable or tenuous the ecological web the greater the necessity for consideration. Acceptance of this concept is evidenced by the protection of wetlands, coastal areas, and virgin woodlands by legislation and ordinances.

Recreational Facilities

Increased census figures, an affluent society, and more leisure time, have increased demand for more recreational facilities. Large numbers of people have migrated towards metropolitan centers because of better employment possibilities; and from the city center to the suburbs for a "better" quality of life, further overcrowding both those locales. This results in a greater need for park space, athletic areas, and other recreational facilities, a trend which apparently will continue into the foreseeable future. All levels of government are aware of this and are attempting to secure more land for additional facilities.

Game lands controlled by the state offer recreation to large numbers of hunters and fishermen each year. Though their use is primarily seasonal, their existence as undeveloped land is very significant.

Camp sites, picnic grounds, and roadside rests are important; however, in terms of location their requirements are not particularly rigid. They also can be altered or moved at reasonable expense.

Such sites can often be developed in conjunction with other construction projects. When completed in this manner they can be built without undue loss of land, within a practical cost framework and limited environmental consequences. Often they may act as a psychological rather than environmental "trade-off".

Golf courses, fee hunting or fishing locations, and winter sports areas represent not only a financial investment, but also a public service. Destruction or displacement, such as major land reduction, should be avoided when possible or compensatory areas developed because of their importance in meeting recreational demand.

Resorts, tourists accommodations, and other similar types of enterprises are generally replaceable, but their intrinsic value to the environment varies with quality, locale, and nature of the facility. Garish strip type development adds little to an area and often serves to reduce surrounding land values. As previously mentioned, inventory items utilized by the public for recreational purposes should be protected and expanded. However, opportunity exists to incorporate additional recreational facilities into new construction projects through imaginative planning.

The consumption rate of raw materials and the projected increase places our natural resource reserves in a perilous position and increases our dependency on other nations with development plans of their own. Efforts are being extended by both government and industry to develop the recycling process as well as methods of curtailing the consumption rate. At this time, however, the overall picture suggests that the surface has barely been scratched and that both funding and research must be intensified.

The concept of "trade-off" is important [6, A-39, 40, *Items 3, 4, 5, 6*]. Ecology is a relationship that is never quite static. It can be disturbed within limits and still readjust. If a change is necessary, what system will result? Could plus factors be introduced? If the new system is potentially as valuable, the trade-off could be considered satisfactory and still allow for the desired development without serious environmental degradation.

This philosophy often meets with opposition from conservationists; however, the question of environmental protection versus public need is a two-sided coin. At one end of the spectrum is "No-Development" and at the other—"Development Without Environmental Concern." Neither absolute must be correct as most environmental decisions are in shades of grey rather than simple black or white. Thus, again, "trade-off" or mitigation must be included in the evaluation process.

Protective features or procedures can be incorporated into the plans and design of a proposed project; this could reduce negative impacts and enhance the environmental value of the facility. It should be emphasized that efforts of this nature should be included in environmental assessment.

This is an appropriate point at which to discuss one of the more nebulous considerations associated with impact evaluation.

Assessing aesthetic quality is extremely subjective except in the most obvious situations. Every construction project will have some effect on this feature with greater impacts generally occurring during construction. Those are temporary, however, and disappear after construction ceases and the scars disappear.

Permanent intrusion, however, is another matter. The destruction of an attractive vista, the extensive excavation of a mountainside where recovery would be difficult and protracted, or the removal of an extremely attractive setting replaced by discordant features without appropriate mitigation should be closely reviewed to see if effects can be softened by changes in location, design, or adequate landscaping efforts.

This area however, is subject to controversy and some limits as to the value of aesthetic quality should be established for a specific project. Every effort should be made to define those limits to be compatible with local conditions and aspirations.

Suggested Reading

Wagner, Richard H., *Environment and Man*, First ed., New York: W. W. Norton & Co., Inc., 1971.

Odum, Eugene P., *Fundamentals of Ecology*, Philadephia: W. B. Saunders Company, 1953.

Geology
  Leet, Donald L. and Judson, Sheldon, *Physical Geology*, Third ed., Englewood Cliffs, N.J.: Prentice-Hall, Inc., 1965.

Vegetation
  The Peterson Series, Boston: Houghton Mifflin Co.
  The Dover Book Series, New York: Dover Publications, Inc.

Wildlife
  Moen, Aaron N., *Wildlife Ecology*, San Francisco: Freeman Press, 1973.
  Booth, *How to Know the Mammals*, Dubuque, Iowa: William C. Brown, 1949.
  The Peterson Series, Boston: Houghton Mifflin Co.
  The Dover Book Series, New York: Dover Publications, Inc.

Natural Resources
  Ciriacy-Wantrup, S. V., *Resource Conservation, Economies and Policies*, Third ed., Berkeley, Calif.: University of California Division of Agricultural Sciences, Agricultural Experiment Station, 1968.

9

SOCIO-ECONOMICS

The number of factors included in this section is large and the spectrum is broad. Several of these are intangible and therefore difficult to pinpoint [2.8, A-23; 4.8, A-36-38]. Projections are also difficult because of the often unpredictable behavior of man himself. Nevertheless they must be addressed by assessing existing conditions and determining potential impacts. Those socio-economic factors should be integrated with physical findings. During NEPA's earlier period there was a tendency to underplay socio-economics and place strong emphasis on physical conditions; however, this philosophy has been altered so that more consideration is necessary.

Traditionally there has been an economic interest in new projects and their effects in terms of dollars. Cost-benefit analyses have been made for some time and have served to justify implementation. However, social effects long ignored have come into prominence as the country has become more aware of the implications and long-term effects on the well being of the various segments of our population. In addition government itself is undertaking actions directly related to social programs, which, by definition, must also be evaluated under NEPA requirements.

Review should cover both temporary and long-range effects. An outline of socio-economic assessment could basically include, depending on the project and its locale, the following factors:

General

    Agriculture production
    Construction activity
    Commerce levels
    Industrial levels
    Public utility availability
    Community form
    Spatial conformation

Direct

    Land value
    Tax base
    Economic base
    Employment levels
    Wage structure
    Income distribution
    Housing supply
    Rent structure
    Local access
    Regional access

Indirect

    Option values
    Distribution of costs
    Distribution of benefits
    External economies
    Regional shifts in output
    Scale economies
    Locational advantages

Social

    Social structure of neighborhoods or communities
    Bisection or stress on neighborhood or community cohesion
    Social structure and character of neighborhoods or communities
    Population trends, local and regional
    Local social institutions

Employment trends
Local governments
Social overhead—
  Health
  Education
  Police
  Fire
  Water, sewage
  Public transit
  Welfare
  Private utilities

There is an increasing emphasis on the inclusion of socio-economic impacts as they relate to major construction projects. Displacement, rate of compensation, land values, disruption, as well as the generation of uncontrolled growth (sprawl) with its ensuing burden on suburban governments to supply adequate service, are just a few possible areas to examine. Central communities cannot sustain a continuing loss of tax revenue as population and business migrate to suburbs and still be able to properly meet their responsibilities.

For example, the factors in the outline should be superimposed on a specific action so as to determine which are to be significantly affected and whether the impact is positive or negative. Employment is important to an area, not only in terms of the actual number of positions available but also the various classifications. Changes in the number or ratio can be significant [4.8, A-37]. With a decline of the employment rate a negative effect on the local economy will result, and vice versa. Always an extremely sensitive area, the economy of the region as well as the surrounding area has to be considered as part of the effect of any proposed project. The spectrum of economic factors being so broad, however, an extremely large project is required to significantly upset regional levels. Conversely, any claim that a project could enhance local economics should be supported by adequate data.

Many factors in this group are strongly interrelated and an effect on one could result in similar impacts on associated features.

The quality of existing public service levels is important in terms of financial responsibility, local goals, and improvement of community life. This topic will be addressed in a discussion of land use implications.

The social structure of a neighborhood, the disruption of that structure by massive displacement, and a change in local "flavor" can all be important [4.8, A-36]. For instance, many older well-established and well-maintained neighborhoods have been set on the path to disintegration by the encroachment of a major highway or another facility of similar magnitude. The loss of residences and the reduction in remaining property values have resulted in wholesale sales or abandonment, and area quality suffers significantly.

There is another problem related to residential and commercial displacement [4.8, A-36]. Although eminent-domain laws have become more liberal, it is increasingly more difficult to locate like housing due to inflation. This obviously could place a hardship on those forced to emigrate. Well-established commercial enterprises likewise suffer when forced to relocate due to loss of their recognized address; unless, of course, the area has already been subject to changes resulting in a steady loss of volume.

This might be discussed under a section entitled *Community Cohesion*. The desire on the part of state and federal governments to maintain local flavor has seriously increased in the last few years.

The other social factors should also be reviewed as they relate to the specific action for consequences. Such situations should be evaluated so that when such impacts are imminent, proper "tradeoffs" can be offered in compensation, if possible.

In developing the scenario of existing conditions excellent data sources include the Chamber of Commerce, church groups, union organizations, neighborhood organizations, and local planning and political agencies.

Both temporary and permanent significant impacts on public services such as police and fire protection, the daily schedule of educational institutions, or utilities including sewage and solid waste disposal, water, sewer, electric, gas, and telephone should be established. Expense, inconvenience, and discomfort can result from interruption of these services. Thus, the need for close liaison among engineers, planners, public agencies, and utilities during the planning process becomes obvious.

It is also patently plain that public health and safety is a prime consideration; if it is significantly threatened by a proposed action, the danger will have to be removed or reduced to an acceptable level.

Most often the socio-economic impacts are longer-ranged and secondary in nature. However, just because their effect is not as dramatic or obvious as physical changes does not reduce their importance as they could shift the environmental balance of the proopsed project.

Suggested Reading

Jarrett, Henry, *Environmental Quality in a Growing Economy*, Baltimore, Md.: Johns Hopkins Press, 1966.

Goldman, Marshall I., *Controlling Pollution—the Economics of a Cleaner America*, Englewood Cliffs, N.J.: Prentice-Hall, 1967.

Herfindahl, O. C. and A. P. Kneese, *Economic Theory of Natural Resources*, Columbus, Ohio: Charles E. Merrill Publishing Company, 1974.

Revelle, R. and H. H. Landsberg, *America's Changing Environment*, Boston: Beacon Press, 1970.

Heilbroner, Robert L., *Understanding Micro-Economics*, Englewood Cliffs, N.J.: Prentice-Hall, 1972.

10

LAND USE

One of the more important facets of new construction and development is the potential impact on land use.

Any action that would tend to generate or retard development can have a significant effect on an area. For instance, the addition of a limited access highway or large sewer collection and disposal system in a sparsely populated region will tend to attract new residents. Often, however, the community may not be ready to receive the influx because other services do not yet have sufficient capacity. Related service levels should be considered in relation to local tax revenues. If they are not adequate to pay for, or act as a base to obtain funding for additional facilities, the project would serve to foster development beyond the financial ability of local government.

Still other undesirable situations include uncontrolled development often called *sprawl,* and *strip development.* Both can be seen in most suburban areas. The former is associated with the bypassing of vacant land located close to a center community capable of serving a large population without additional investment. The possibility of a proposed plan or project to generate these conditions should be closely examined.

On the other hand, if a project causes the population to leave an area, which, as a result, loses its tax base and becomes subject to diminishing property values, the project will also have a negative effect as it would accelerate both the loss of people and tax monies.

These basic concepts should be considered during evaluation to keep the proposed project in perspective. It is important for local government to have a set of "tools" available to protect their community's open space, buffer zones, as well as the goals and aspirations established by local citizens. These could include:

1. Comprehensive plans,
2. Land use or zoning plans,
3. Ordinances,
4. Building codes.

Concurrence with and enforcement of these will serve well to prevent uncontrolled, undesirable growth, and allow for orderly development.

One important procedure is to determine the planned land use for property to be taken for a project. For instance, a highway may take several acres of good agricultural land, which would have a negative impact. However, if land use plans indicate that the area is designated for mid- or high-density residential and commercial development, would the loss of that land by the highway be any more detrimental than by residential development? This is not to be interpreted that this rationale can be universally applied as a basis for condoning 'free-wheeling" development; however, the situation should be examined and if necessary corrected by local officials.

The basis for good land use includes a knowledge of the environmental inventory. Without this information it is difficult to address optimum use on a logical basis as there is little concept of what is to be protected or enhanced, which must be understood to determine the best use for undeveloped acreage.

Again, in terms of impact assessment, local land use plans should be reviewed during the process to determine whether the project will conform with local plans or not.

Suggested Reading

Soils
Bartelli, Lindo J. et al., ed., *Soil Surveys and Land Use Planning*, Chapter 1, pp. 1-7; Chapter 2, pp. 8-14; Chapter 3, pp. 15-30; Chapter 2, pp. 42-59; and Plates 1-17, pp. 181-196. Madison, Wisc.: Soil Science Society of America, 1966.

Maps, Surveys
Way, Douglas S., *Terrain Analysis*, Chapter 4, pp. 53-65, Stroudsburg, Pa.: Dowden, Hutchinson & Ross, Inc., 1973.

Introduction: Environmental Planning
U.S. National Science Foundation, RANN. *Environment: A New Focus for Land Use Planning*, ed. by Donald M. McAllister (NSF/RA/E-74-001, Oct. 1973). Chapter 8, pp. 183-215; Chapter 10, pp. 233 247; Chapter 12, pp. 267-278; Chapter 14, pp. 291-300.

Land Classification
Lacate, Douglas S., "A Review of Landtype Classification and Mapping," Land Economics 37(3): 271-278 (August 1961).

Land Capability Approaches
Clawson, Marion, & Charles L. Stewart. *Land Use Information* (Baltimore: Johns Hopkins for Resources for the Future, 1965). Chapter I, pp. 1-4; Chapter II, pp. 11-21; Chapter VII, pp. 142-153.

Hills, G. A., *The Ecological Basis for Land Use Planning*. Research Report No. 46, Ontario Dept. of Lands & Forests, Research Branch, Dec. 1961. Chapter I, pp. 1-5; Chapter II, pp. 16-30, 46-49; Chapter VIII, pp. 143-152; Chapter IX, pp. 153-180.

Administrative Agency Approaches
Soil Conservation Society of America, *Planning & Zoning for Better Resource Use*. Prepared by the Land Use Planning Committee, 1971.

Landscape Resource Analysis
McHarg, Ian L., *Design with Nature*, Garden City, N.Y.: Natural History Press, 1969, pp. 31-41, 103-115, 127-151.

11

HISTORIC AND ARCHEOLOGICAL SITES

Although not part of the natural inventory many of these historic structures, representing earlier human efforts have become important national assets. While they are on occasion merely objects of sentimental interest, there is no question that many are irreplaceable in terms of age and craftsmanship.

Archeological resources have also been recognized and more actively protected during the last decade. Previously they were considered relatively unimportant, except to a limited number of dedicated scientists.

By definition, historic resources include sites, districts, structures, objects, or other evidences of human activities that represent facets of the history of nation, state, or locality; places where significant historical or unusual events occurred even though no evidence of the event remains; or places associated with a personality important in history.

Man-made structures may also be designated as historic if they are particularly representative of their class or period, represent achievements—in architecture, engineering, design, or scientific research and development—which possess artistic merit, or represent the work of a master [2.9, A-23; 4.9, A-38].[1]

Archeological resources are simply objects and/or areas made or modified by man which can be found at occupation sites, hunting areas, ceremonial gathering spots, or burial grounds which are now below the ground surface or at the surface.

Legislation protecting these features dates back to the Antiquity Act of 1906 (Public Law 59-209, 34 STAT. 225; 16 U.S.C. 431-433) which sets forth the basic tenet that the federal government should take the lead in working for the preservation and protection of the nation's historic and archaeological resources.

The Historic Sites Act of 1935 (Public Law 74-292, 49 STAT. 666; 16 U.S.C. 461-467) repeated this principal and directed the National Park Service to collect, preserve, and restore such features.

The Historic Preservation Act of 1966 (Public Law 89-665, 80 STAT. 915; 16 U.S.C. 470), however, was much broader in scope. It authorized the Secretary of the Interior to expand and maintain a national register of districts, sites, buildings, structures, and objects significant in American history, architecture, archeology, and culture. States were also granted funds to undertake state-wide surveys in order to nominate sites for the national register. Matching funds were also provided to help acquire and preserve historic sites. In addition, state agencies and local societies have set up separate lists.

In the evaluation process the national register, state museum, and local groups should be contacted to insure that sites on the registers and/or those being considered are located and the impact of the proposed project evaluated.

Field surveys may produce other potential structures. Age, quality, type of construction, and the number of similar units in the immediate area should be considered.

While aged structures merit protection for historical, cultural, and educational reasons, levels of protection for registered sites should be kept in perspective. Old forts, battlefields, and ghost towns have, on occasion, been the basis for rather garish tourist attractions. These commercial enterprises can be operated in such a manner as to make them an environmental asset; conversely, they can blemish a local community. Therefore, not only the specific

---

[1] Wright, Russell, "Techniques for Incorporating Historic Preservation Objectives Into the Highway Planning Process," p. xi, Prepared for the National Trust for Historic Preservation, Washington, D.C., December 1972.

item, but also the associated activities should be considered. It is possible, lacking other alternatives, to move a valuable structure to another location. For instance, covered bridges, revered relics of an earlier technology, can be spared and moved to serve as entrances to parks; however, provisions should be made for proper maintenance to insure that they do not become eyesores.

Cemeteries should be protected not only for religious considerations but because they form a valuable link with the past in the form of tombstones and inscriptions. They also serve in preserving open space in areas otherwise completely built up. Several states and communities have established ordinances against the destruction or movement of interment sites without specific approvals. These regulations can be checked through local officials.

In the evaluation of historic items, change in project location, possible movement, or destruction, are the only construction alternatives. Thus the options are limited and should be properly reviewed.

Archeological sites, generally relatively limited areas, are an important link to the past with a research value directly dependent on their size, content, and age. They also serve as an educational aid allowing those interested in archeology to forward this interest for the benefit of a society that is often ignorant of its ancient culture.

Because of increased interest and because expertise in this field is limited, contact with state archeologist offices is wise as they may be prepared to report on site or area potential.

Suggested Reading

Wright, Russell, "Techniques for Incorporating Historic Preservation Objectives into the Highway Planning Process," prepared for the National Trust for Historic Preservation, Washington, D.C., December 1972.
National Register of Historic Places

12

GRAPHIC AIDS—
REMOTE SENSING

Clearly presenting environmental impacts can be important. Often these conditions can be demonstrated graphically showing quantitative as well as qualitative limits of effect.

Both regional and local maps should be included. The alternatives should be shown and information such as the following should be shown:

    Air Quality monitoring sites
    Acoustic monitoring sites
    Water Quality monitoring sites
    Historical/Archaeological site locations
    Agricultural land (Levels I and II)
    Wildlife areas
    Existing and proposed land-use plans
    Recreation areas

as well as other significant features in the study area that might be affected by the project.

Appropriate scales will be determined by the size of the area to be portrayed.
Other graphic aids could include:

1. Aerial photography
2. Overlays
3. Photomosaics

Aids have been used that are based on aerial photography and maps. New photography and mapping is most desirable in this type of work as it will include the latest features and development. However, it is often not practical either for reasons of time or money to purchase new photos or mapping. Under those conditions use the most practical detailed map available and U. S. Agricultural or even ERTS photography. Both are available for minimal charges.

AERIAL PHOTOGRAPHY

If new pictures are obtained, they should be vertical photography. These are pictorial and can serve as the basis for generating accurate measurements by photogrammetry. If mapping is desired, the photographs are to be taken with a distortion-free aerial camera at a negative (photographic) scale, depending upon the size of the area involved and the scale of the map desired. The photography can also be taken in color or infra-red for greater photo interpretation capability. See photograph on next page.

OVERLAYS

Using overlays for the presentation of data is a proven and effective method of demonstration, because they are versatile and have strong pictorial capabilities. The information to be shown lends itself especially to this system.

Overlays of clear film or mylar can be made with any of the large number of materials available. Type should be oversized so that if reduced for printing it will be legible. The overlays should be registered with the base map.

Vertical Aerial Photograph (Scale 1" = 1667)
(Courtesy: Berger Associates, Columbus, Ohio)

Photomosaic
(Courtesy: Berger Associates, Columbus, Ohio)

PHOTOMOSAICS

A photomosaic consists of several aerial photographs of the same series skilfully matched so that the result appears to be one large photograph. Photomosaics are recommended for visual presentation because they can be produced reasonably at any desired size and scale. They show detail; and can demonstrate quantity and position, thus making an excellent base for overlays. See photograph opposite page.

The aids described above can be developed into a presentation to locate as well as delineate potential losses, gains, and physical impacts at relatively low cost.

REMOTE SENSING

In the last several years there have been significant gains in the field of *remote sensing,* which is the collection of information from such mediums as aerial photography, telephotos, and magnetometer data. Resulting information has proved to be not only voluminous but extremely accurate.

Many agencies and university departments are engaged in sensing activities which can serve as data sources. In fact entire systems have been established that serve not only to develop inventory data, but also serve as monitors indicating excessive pollution levels.

These activities and the subsequent data can be effectively used in environmental impact evaluation.

Suggested Reading

American Society of Photogrammetry
  Photogrammetric Engineering (monthly journal).
  Manual of Photogrammetry, Third ed., 1966.
  Manual of Photographic Interpretation, 1960.
  Manual of Color Aerial Photography, 1968.
  Manual of Remote Sensing (1974).

COLWELL, ROBERT N., "Remote Sensing of Natural Resources," *Scientific American,* Vol. 218(1), New York: Scientific American, Inc., January 1968.

Environmental Research Institute of Michigan (ERIM)
(Box 618, Ann Arbor, Michigan 48107; formerly Willow Run Laboratories, Univ. of Michigan)

Proceedings of the Symposia on Remote Sensing of Environment
First, February 1962.
Second, October 1962.
Third, October 1964.
Fourth, April 1966.
Fifth, April 1968.
Sixth, October 1969.
Seventh, May 1971.
Eighth, October 1972.
Ninth, April 1974.

LOWMAN, PAUL D., Jr., "The Earth from Orbit," *National Geographic,* Vol. 130(5), November 1966.

Scherz, James P. and Alan R. Stevens, *An Introduction to Remote Sensing for Environmental Monitoring* (Remote Sensing Program, Institute for Environmental Studies, The Univ. of Wisconsin, August 1970).

WEAVER, KENNETH F., "Remote Sensing: New Eyes to See the World," *National Geographic,* Vol. 135(1), January 1969.

13

THE DRAFT ENVIRONMENTAL STATEMENT

This is the document which will contain the information generated during the assessment period. It is the formal report on environmental aspects of a proposed action.

The format is established and stylized by the regulations of the lead agency. However, any assessment involves a description of existing environmental conditions, the placing of the proposed project and its alternatives in the setting and presenting the effects of implementing the various options, and comparing the various positive and negative impacts. This should be done so that along with other input such as engineering ramifications, costs, and usage, the decision-makers can weigh the net positive and negative results and come to a final conclusion substantiated by pertinent data.

To present this information in a logical, understandable manner the following basic format should be utilized within the specified requirements of the agency guidelines.

    I. *Table of Contents.*
        A. By Section
        B. List of Figures

C. List of Tables

II. *Summary of Report.*

III. *Project Surroundings.*
Regional
Local

This is to include such information as distance factors to metropolitan areas, geographic and economic data, description of regional background, and other information, such as topography and river basin relationships, all of which portrays general conditions of the area.

The immediate locale is also to be described in the same manner but in greater detail.

IV. *A Chronological History of the Project.* This would include the planning process, previous studies and reports. The problem to be addressed and the objectives to be served should also be included. Alternatives previously studied are to be presented and rationale when deemed unfeasible.

Previous agency and public coordination efforts and reactions should be addressed and documented. Ongoing efforts should also be included so that project status is current.

V. *Existing Environmental Conditions.* This can encompass several different factors depending on project locale, however, they generally include descriptions of the following considerations:

A. Climate,
B. Air Quality,
C. Acoustic Levels,
D. Water Quality,
E. Ground Water/Aquifiers,
F. Soils/Geology,
G. Terrestrial Ecology,
H. Aquatic Ecology,
I. Land Use,
J. Socio-Economic Factors,
K. Aesthetics,
L. Historic/Archeological Sites,
M. Description of Public Services

VI. *Alternatives.*
   A. Construction
   B. Non-Construction
   C. No Action

These should be described in detail so that attributable effects can be easily associated with each option. Comparative costs are generally placed in this section.

VII. *Environmental Impacts.* This section ideally should be structured so that the factors can be compared with those listed under Existing Conditions. There can be, however, additional factors such as displacement and destruction numbers. Also, depending on Agency formats, there may be a different category outline that more efficiently lists the impacts related to their primary type of project.

Significant positive and negative impacts should be specified and the effects of each option should be indicated by factor for easy comparison. Both temporary and permanent impacts must be included.

VIII. *Unavoidable Adverse Impacts.* This section should summarize significant negative impacts, both temporary and permanent, which would be associated with the proposed project.

IX. *Measures Under Consideration to Minimize Unavoidable Negative Environmental Effects.* This section should list those measures that could be adopted to reduce negative impacts and possibly improve existing conditions. This is where "trade-offs" are incorporated.

X. *The Relationship Between Local Short Term Uses of Man's Environment and the Maintenance and Enhancement of Long Term Productivity.* This section should discuss the short term versus long term effects of the proposed action. It should include long range impacts on both physical and socio-economic factors.

XI. *Any Irreversible and Irretrievable Commitment of Resources which would be Involved in the Proposed Action were it Implemented.* The purpose of this section is to list the type of resources to be utilized that could not, in effect, be retrievable. They could include natural resources, such as land, construction materials, energy man-power, and financial investment, among others.

XII. *Appendix.* Supporting data, too cumbersome for the statement proper, should be placed in this section as well as official project correspondence, such as agency responses.

The first submittal is designated as the draft. It is this text that is reviewed by all concerned governmental agencies and the public, and used as an information source for the public hearing. As it is subject to criticism as well as requests for additional information and clarification, it should be written as thoroughly as possible to assure a minimum of additional input. This not only reduces further research, but lends greater credibility to the proposed project.

All feasible alternatives including "no-construction" must be addressed.

Items (IV) and (V) of 102(c) should not be treated lightly. They involve questions relating to the entire environmental spectrum.

Finally, the statement should include those recommendations that could reduce adverse environmental impacts. This section is extremely important as it demonstrates the concern of project proponents for the environment and introduces possible "trade-offs."

Submittal should be on 8½ x 11" paper with the formal cover and a very brief summary preceding the body. It is wise to double space the draft so that there is room for comments by reviewers.

To repeat, the writing style should be formal, as the EIS is a technical document, and extensive use of embellishing adjectives tends to discredit a scientific effort.

Also, maps, illustrations, photographs, and charts should be included to support or demonstrate the body of the text. If the assessment has determined that there will be no significant impacts, then a *negative declaration* accompanies a written report of the environmental assessment.

The draft statement should contain, in the appendix, all official correspondence related to the project and a record of coordination efforts. This is to include agency and public communications.

The finished product will be signed off by the lead agency and circulated to other interested agencies, organizations, and individuals for review.

In terms of assessment costs and man-hour distribution, the following information may assist in evaluation planning.

1. In completing an average draft and final statement, approximately *80–85 percent* of the effort will be devoted to the draft, and 15–20 percent to the final.

2. Approximately 45 to 50 percent of draft EIS time will be ex-

pended in establishing existing conditions. This does not include *extensive* monitoring for ambient air quality, but does include establishing existing acoustic levels.

3. Project/study management can require up to 5% to 8% of total allotted man-hours. This is necessary to insure coordination, continuity, and timely delivery.

4. Specialized expertise is available at most colleges and universities as well as consulting firms.

5. Man-hours for attendance at public meetings and the public hearings should be included.

6. Gathering existing data from agencies and universities can significantly reduce expenditures and should be done before field work begins.

7. Monitoring equipment for both air quality and acoustic levels can be leased.

8. Miscellaneous study expenses can include:
 (a) travel,
 (b) lodging and subsistence,
 (c) telephone,
 (d) drafting hours and supplies,
 (e) printing,
 (f) postage.

14

PUBLIC MEETINGS AND HEARINGS

The public hearing can be held, depending on agency procedures, thirty days after the publishing and distribution of the Environmental Impact Statement; however, because of increasing public interest, the more time between EIS distribution and the hearing, the greater time for review and the less likelihood of there being complaints because of insufficient time to become familiar with the EIS.

In those undertakings that tend to be controversial, a series of public meetings should be initiated. These serve to familiarize the local citizens with the project and ask for assistance in order to create a more effective program or facility. These meetings do not have to be *totally* official, but can act as information briefings and as "sounding boards" for public opinion and contributions. Often, if strong opposition groups arise it is important to have separate meetings with their representatives to offer additional clarification and to obtain additional input which could literally change the project to satisfy objections and goals.

Although much valuable data can be obtained from citizens

geographically or in terms of interest, this type of approach can create its very own set of problems. At this stage the project is never perfectly detailed, yet those that might be potentially affected become alarmed and demand specifics that are not yet available. This early exposure, if not properly handled can create a "credibility gap" and generate opposition even before any real formulation has been completed. It must be made plain that this type of meeting is for obtaining local input that will aid the decision-makers in determining whether the project is feasible and compatible with local policy. This should then assist the design group in creating a project that will best serve public interests. Printed material explaining the project should be distributed if possible.

Generally the Public Hearing is required by law. It is chaired and has an established procedure. The official purpose of this gathering is a public review of a proposed project, the alternatives, the financial aspects, and the full scope of environmental consequences.

At many hearings, depending on the agency, the proposal is presented, and public officials, organizations, and private citizens can offer in testimony their opinions, whether supportive or not. Questions or requests can be presented. This is duly transcribed and becomes part of public record and the final EIS.

Preparation for all public sessions should be well thought out and complete. They are not intended to serve as sales meetings; however, a clear, concise, knowledgeable, unbiased presentation with a strong emphasis on graphics is necessary. The greatest majority of those attending are not professional engineers or planners, and the presentation, including the graphics, should be readily understandable.

The public has become skeptical of new projects. This results not only in active opposition, but in demand for design details which generally have not been fully determined at the hearing stage. It is important to have those attending understand that the project is merely being offered as a possibility. It has not been approved or funded and their comments could be part of the input towards the final decision.

Most projects of any magnitude generate public interest and opposition, the latter running the spectrum from irate individuals to highly organized groups to Congressional representatives. Certain basic interest patterns, however, are becoming evident.

1. Localized opposition by those directly affected by the project geography.
2. Conservationists dedicated to the preservation of wilderness and open space.
3. Vested interest groups.
4. Activists opposed to major construction.

This interest/opposition occurs for many reasons, ranging from altruistic environmental considerations to being personally and directly affected by the proposed action or project. One obvious cause is the extended period between planning and implementation. Another is the growing public awareness and interest in governmental actions. Philosophies, growth, and needs are changing much too rapidly to find a passive public. The population is too well educated and too knowledgeable to accept twenty-year old information as credible. Therefore good judgment is required in choosing the material to be presented and its validity. Updating may be necessary in view of the long period between planning and implementation.

Public hearings have, in many cases, acted as "sounding boards" for project opposition, although this is included in the intent of the system. Those conducting the forum should keep control, or amid the noise, applause, and other distractions not conducive to generating a true index of public reaction the true purpose of the public hearing will be lost.

Suggested Reading

The following have completed several studies and published articles on their approach to fostering public participation:

U.S. Army Corps of Engineers
Environmental Protection Agency
Soil Conservation Service

15

THE FINAL STATEMENT

Before writing the final EIS, comments offered at the public hearing and responses from official agencies have to be reviewed. Pertinent suggestions, undetected flaws, and the lack of information are often brought to focus during public testimony. Agency response can be to either "sign off" on the project, indicating that they have no concern for the effects of the project on their area of responsibility, or requesting additional data, when they feel the draft is deficient.

Legitimate inquiries should be addressed, even at the expense of additional investigation, as they represent unanswered questions of both the public and responsible agencies. This additional information, all correspondence, testimony, and the recommended action are added to the draft as well as modifications to the projects, and a copy of the draft EIS. This represents the final Environmental Impact Statement, which is reviewed for action approval.

It is wise to annotate in print the questions raised by agencies and the public in official correspondence and the hearing, with the paragraph and page number, so that the final EIS will reflect this information.

In some cases associated submittals are required for approval. These could include a permit from EPA, if there is to be a discharge into a navigable stream, a permit from the U.S. Army Corps. of Engineers, or a 4(f) Statement (Chapter 4), if public land is to be taken.

The total package, the Final EIS signed by the lead agency, is then redistributed and filed with the Council of Environmental Quality.

Final project approval rests with the head of the lead agency.

16

AUTHOR'S COMMENTS

It is the author's hope that this manual has presented to the reader the rationale and procedures of environmental impact evaluation as well as the NEPA process. At this point in time the state of the art is still developing and many diverse approaches to both data collection and projections are being promulgated. I believe, however, that because of the large number of factors and the almost infinite number of interrelationships, both simple procedures and complicated computer matrices should be carefully scrutinized before acceptance as optimum solutions. Expertise and experience are important in impact projections and over-simplification or over-complication of the process can lead to insensitive conclusions.

Classes and seminars on environmental impact evaluation are being offered through several sources. Hopefully, with research, experience, and improved techniques, the process will become more sophisticated and efficient, concentrating more on the real and significant issues.

In the interim those assigned to responsibility in the NEPA process should be informed as to the latest agency guidelines and memoranda, as well as new methodology.

Finally, in the preparation of an EIS, it is extremely important to keep the statement in perspective with the project. A $100,000.00 report on a $50,000.00 facility would certainly be indicative of an "overkill." Thus, judgment should be exercised in the development of the EIS.

APPENDIX

1

National Environmental Policy Act of 1969

NATIONAL ENVIRONMENTAL POLICY ACT OF 1969
91st Congress, S. 1075
January 1, 1970

An Act

83 STAT. 852

To establish a national policy for the environment, to provide for the establishment of a Council on Environmental Quality, and for other purposes.

National Environmental Policy Act of 1969.

Be it enacted by the Senate and House of Representatives of the United States of America in Congress assembled, That this Act may be cited as the "National Environmental Policy Act of 1969".

PURPOSE

SEC. 2. The purposes of this Act are: To declare a national policy which will encourage productive and enjoyable harmony between man and his environment; to promote efforts which will prevent or eliminate damage to the environment and biosphere and stimulate the health and welfare of man; to enrich the understanding of the ecological systems and natural resources important to the Nation; and to establish a Council on Environmental Quality.

TITLE I

DECLARATION OF NATIONAL ENVIRONMENTAL POLICY

Policies and goals.

SEC. 101. (a) The Congress, recognizing the profound impact of man's activity on the interrelations of all components of the natural environment, particularly the profound influences of population growth, high-density urbanization, industrial expansion, resource exploitation, and new and expanding technological advances and recognizing further the critical importance of restoring and maintaining environmental quality to the overall welfare and development of man, declares that it is the continuing policy of the Federal Government, in cooperation with State and local governments, and other concerned public and private organizations, to use all practicable means and measures, including financial and technical assistance, in a manner calculated to foster and promote the general welfare, to create and maintain conditions under which man and nature can exist in productive harmony, and fulfill the social, economic, and other requirements of present and future generations of Americans.

(b) In order to carry out the policy set forth in this Act, it is the continuing responsibility of the Federal Government to use all practicable means, consistent with other essential considerations of national policy, to improve and coordinate Federal plans, functions, programs, and resources to the end that the Nation may—

(1) fulfill the responsibilities of each generation as trustee of the environment for succeeding generations;

(2) assure for all Americans safe, healthful, productive, and esthetically and culturally pleasing surroundings;

(3) attain the widest range of beneficial uses of the environment without degradation, risk to health or safety, or other undesirable and unintended consequences;

(4) preserve important historic, cultural, and natural aspects of our national heritage, and maintain, wherever possible, an environment which supports diversity and variety of individual choice;

(5) achieve a balance between population and resource use which will permit high standards of living and a wide sharing of life's amenities; and

(6) enhance the quality of renewable resources and approach the maximum attainable recycling of depletable resources.

(c) The Congress recognizes that each person should enjoy a healthful environment and that each person has a responsibility to contribute to the preservation and enhancement of the environment.

Administration.

SEC. 102. The Congress authorizes and directs that, to the fullest extent possible: (1) the policies, regulations, and public laws of the United States shall be interpreted and administered in accordance with the policies set forth in this Act, and (2) all agencies of the Federal Government shall—

(A) utilize a systematic, interdisciplinary approach which will insure the integrated use of the natural and social sciences and the environmental design arts in planning and in decisionmaking which may have an impact on man's environment;

(B) identify and develop methods and procedures, in consultation with the Council on Environmental Quality established by title II of this Act, which will insure that presently unquantified environmental amenities and values may be given appropriate consideration in decisionmaking along with economic and technical considerations;

(C) include in every recommendation or report on proposals for legislation and other major Federal actions significantly affecting the quality of the human environment, a detailed statement by the responsible official on—

   (i) the environmental impact of the proposed action,

   (ii) any adverse environmental effects which cannot be avoided should the proposal be implemented,

   (iii) alternatives to the proposed action,

   (iv) the relationship between local short-term uses of man's environment and the maintenance and enhancement of long-term productivity, and

   (v) any irreversible and irretrievable commitments of resources which would be involved in the proposed action should it be implemented.

Prior to making any detailed statement, the responsible Federal official shall consult with and obtain the comments of any Federal agency which has jurisdiction by law or special expertise with respect to any environmental impact involved. Copies of such statement and the comments and views of the appropriate Federal, State, and local agencies, which are authorized to develop and enforce environmental standards, shall be made available to the President, the Council on Environmental Quality and to the public as provided by section 552 of title 5, United States Code, and shall accompany the proposal through the existing agency review processes;

(D) study, develop, and describe appropriate alternatives to to recommended courses of action in any proposal which involves unresolved conflicts concerning alternative uses of available resources;

(E) recognize the worldwide and long-range character of environmental problems and, where consistent with the foreign policy of the United States, lend appropriate support to initiatives, resolutions, and programs designed to maximize international cooperation in anticipating and preventing a decline in the quality of mankind's world environment;

(F) make available to States, counties, municipalities, institutions, and individuals, advice and information useful in restoring, maintaining, and enhancing the quality of the environment;

(G) initiate and utilize ecological information in the planning and development of resource-oriented projects; and

(H) assist the Council on Environmental Quality established by title II of this Act.

Review.

SEC. 103. All agencies of the Federal Government shall review their present statutory authority, administrative regulations, and current policies and procedures for the purpose of determining whether there are any deficiencies or inconsistencies therein which prohibit full compliance with the purposes and provisions of this Act and shall propose to the President not later than July 1, 1971, such measures as may be necessary to bring their authority and policies into conformity with the intent, purposes, and procedures set forth in this Act.

SEC. 104. Nothing in Section 102 or 103 shall in any way affect the specific statutory obligations of any Federal agency (1) to comply with criteria or standards of environmental quality, (2) to coordinate or consult with any other Federal or State agency, or (3) to act, or refrain from acting contingent upon the recommendations or certification of any other Federal or State agency.

SEC. 105. The policies and goals set forth in this Act are supplementary to those set forth in existing authorizations of Federal agencies.

TITLE II

COUNCIL ON ENVIRONMENTAL QUALITY

Report to Congress.

SEC. 201. The President shall transmit to the Congress annually beginning July 1, 1970, an Environmental Quality Report (hereinafter referred to as the "report") which shall set forth (1) the status and condition of the major natural, manmade, or altered environmental classes of the Nation, including, but not limited to, the air, the aquatic, including marine, estuarine, and fresh water, and the terrestrial environment, including, but not limited to, the forest, dryland, wetland, range, urban, suburban and rural environment; (2) current and foreseeable trends in the quality, management and utilization of such environments and the effects of those trends on the social, economic, and other requirements of the Nation; (3) the adequacy of available natural resources for fulfilling human and economic requirements of the Nation in the light of expected population pressures; (4) a review of the programs and activities (including regulatory activities) of the Federal Government, the State and local governments, and nongovernmental entities or individuals, with particular reference to their effect on the environment and on the conservation, development and utilization of natural resources; and (5) a program for remedying the deficiencies of existing programs and activities, together with recommendations for legislation.

Council on Environmental Quality.

SEC. 202. There is created in the Executive Office of the President a Council on Environmental Quality (hereinafter referred to as the "Council"). The Council shall be composed of three members who shall be appointed by the President to serve at his pleasure, by and with the advice and consent of the Senate. The President shall designate one of the members of the Council to serve as Chairman. Each member shall be a person who, as a result of his training, experience, and attainments, is exceptionally well qualified to analyze and interpret environmental trends and information of all kinds; to appraise programs and activities of the Federal Government in the light of the policy set forth in title I of this Act; to be conscious of and responsive to the scientific, economic, social, esthetic, and cultural needs and interests of the Nation; and to formulate and recommend national policies to promote the improvement of the quality of the environment.

SEC. 203. The Council may employ such officers and employees as may be necessary to carry out its functions under this Act. In addition, the Council may employ and fix the compensation of such experts and consultants as may be necessary for the carrying out of its functions under this Act, in accordance wtih section 3109 of title 5, United States Code (but without regard to the last sentence thereof).

80 Stat. 416.
Duties and functions.

SEC. 204. It shall be the duty and function of the Council—

(1) to assist and advise the President in the preparation of the Environmental Quality Report required by section 201;

(2) to gather timely and authoritative information concerning the conditions and trends in the quality of the environment both current and prospective, to analyze and interpret such information for the purpose of determining whether such conditions and trends are interfering, or are likely to interfere, with the achievement of the policy set forth in title I of this Act, and to compile and submit to the President studies relating to such conditions and trends;

(3) to review and appraise the various programs and activities of the Federal Government in the light of the policy set forth in title I of this Act for the purpose of determining the extent to which such programs and activities are contributing to the achievement of such policy, and to make recommendations to the President with respect thereto;

(4) to develop and recommend to the President national policies to foster and promote the improvement of environmental quality to meet the conservation, social, economic, health, and other requirements and goals of the Nation;

(5) to conduct investigations, studies, surveys, research, and analyses relating to ecological systems and environmental quality;

(6) to document and define changes in the natural environment, including the plant and animal systems, and to accumulate necessary data and other information for a continuing analysis of these changes or trends and an interpretation of their underlying causes;

(7) to report at least once each year to the President on the state and condition of the environment; and

(8) to make and furnish such studies, reports thereon, and recommendations with respect to matters of policy and legislation as the President may request.

SEC. 205. In exercising its powers, functions, and duties under this Act, the Council shall—

34 F. R. 8693.

(1) consult with the Citizens' Advisory Committee on Environmental Quality established by Execu-

tive Order numbered 11472, dated May 29, 1969, and with such representatives of science, industry, agriculture, labor, conservation organizations, State and local governments and other groups, as it deems advisable; and

(2) utilize, to the fullest extent possible, the services, facilities, and information (including statistical information) of public and private agencies and organizations, and individuals, in order that duplication of effort and expense may be avoided, thus assuring that the Council's activities will not unnecessarily overlap or conflict with similar activities authorized by law and performed by established agencies.

Tenure and compensation. 80 Stat. 460, 461.

SEC. 206. Members of the Council shall serve full time and the Chairman of the Council shall be compensated at the rate provided for Level II of the Executive Schedule Pay Rates (5 U.S.C. 5313). The other members of the Council shall be compensated at the rate provided for Level IV of the Executive Schedule Pay Rates (5 U.S.C. 5315).

81 Stat. 638. Appropriations.

SEC. 207. There are authorized to be appropriated to carry out the provisions of this Act not to exceed $300,000 for fiscal year 1970, $700,000 for fiscal year 1971, and $1,000,000 for each fiscal year thereafter.

Approved January 1, 1970.

LEGISLATIVE HISTORY:
HOUSE REPORTS: No. 91-378, 91-378, pt. 2, accompanying H. R. 12549
        (Comm. on Merchant Marine & Fisheries) and 91-765
        (Comm. of Conference).
SENATE REPORT No. 91-296 (Comm. or Interior & Insular Affairs).
CONGRESSIONAL RECORD, Vol. 115 (1969):
    July 10: Considered and passed Senate.
    Sept. 23: Considered and passed House, amended, in lieu of H. R. 12549
    Oct.   8: Senate disagreed to House amendments; agreed to conference.
    Dec. 20: Senate agreed to conference report.
    Dec. 22: House agreed to conference report.

APPENDIX 2

Council on Environmental Quality Guidelines

COUNCIL ON ENVIRONMENTAL QUALITY

PREPARATION OF ENVIRONMENTAL IMPACT STATEMENTS

Guidelines

RULES AND REGULATIONS

Title 40—Protection of the Environment

CHAPTER V—COUNCIL ON ENVIRONMENTAL QUALITY

PART 1500—PREPARATION OF ENVIRONMENTAL IMPACT STATEMENTS: GUIDELINES

On May 2, 1973, the Council of Environmental Quality published in the FEDERAL REGISTER, for public comment, a proposed revision of its guidelines for the preparation of environmental impact statements. Pursuant to the National Environmental Policy Act (P.L. 91–190, 42 U.S.C. 4321 et seq.) and Executive Order 11514 (35 FR 4247) all Federal departments, agencies, and establishments are required to prepare such statements in connection with their proposals for legislation and other major Federal actions significantly affecting the quality of the human environment. The authority for the Council's guidelines is set forth below in § 1500.1. The specific policies to be implemented by the guidelines are set forth below in § 1500.2.

The Council received numerous comments on its proposed guidelines from environmental groups, Federal, State, and local agencies, industry, and private individuals. Two general themes were presented in the majority of the comments. First, the Council should increase the opportunity for public involvement in the impact statement process. Second, the Council should provide more detailed guidance on the responsibilities of Federal

This article appeared in *The Federal Register,* Washington, D.C., Vol 38 (147), Part II, August 1, 1973.

agencies in light of recent court decisions interpreting the Act. The proposed guidelines have been revised in light of these specific comments relating to these general themes, as well as other comments received, and are now being issued in final form.

The guidelines will appear in the Code of Federal Regulations in Title 40, Chapter V, at Part 1500. They are being codified, in part, because they affect State and local governmental agencies, environmental groups, industry, and private individuals, in addition to Federal agencies, to which they are specifically directed, and the resultant need to make them widely and readily available.

Sec.
1500.1 Purpose and authority.
1500.2 Policy.
1500.3 Agency and OMB procedures.
1500.4 Federal agencies included; effect of the act on existing agency mandates.
1500.5 Types of actions covered by the act.
1500.6 Identifying major actions significantly affecting the environment
1500.7 Preparing draft environmental statements; public hearings.
1500.8 Content of environmental statements.
1500.9 Review of draft environmental statements by Federal, Federal-State, and local agencies and by the public
1500.10 Preparation and circulation of final environmental statements.
1500.11 Transmittal of statements to the Council; minimum periods for review; requests by the Council.
1500.12 Legislative actions.
1500.13 Application of section 102(2)(C) procedure to existing projects and programs.
1500.14 Supplementary guidelines; evaluation of procedures.

Sec.
Appendix I Summary to accompany draft and final statements.
Appendix II Areas of environmental impact and Federal agencies and Federal State agencies with jurisdiction by law or special expertise to comment thereon.
Appendix III Offices within Federal agencies and Federal-State agencies for information regarding the agencies' NEPA activities and for receiving other agencies' impact statements for which comments are requested.
Appendix IV State and local agency review of impact statements.

AUTHORITY: National Environmental Act (P.L. 91-190, 42 U.S.C. 4321 et seq.) and Executive Order 11514.

§ 1500.1 Purpose and authority.

(a) This directive provides guidelines to Federal departments, agencies, and establishments for preparing detailed environmental statements on proposals for legislation and other major Federal actions significantly affecting the quality of the human environment as required by section 102(2)(C) of the National Environmental Policy Act (P.L. 91-190, 42 U.S.C. 4321 et. seq.) (hereafter "the Act"). Underlying the preparation of such environmental statements is the mandate of both the Act and Executive Order 11514 (35 FR 4247) of March 5, 1970, that all Federal agencies, to the fullest extent possible, direct their policies, plans and programs to protect and enhance environmental quality. Agencies are required to view their actions in a manner calculated to encourage productive and enjoyable harmony between man and his environment, to promote efforts preventing or eliminating damage to the environment and biosphere and stimulating the health and welfare of man, and to enrich the understanding of the ecological systems and natural resources important to the Nation. The objective of section 102(2)(C) of the Act and of these guidelines is to assist agencies in implementing these policies. This requires agencies to build into their decisionmaking process, beginning at the earliest possible point, an appropriate and careful consideration of the environmental aspects of proposed action in order that adverse environmental effects may be avoided or minimized and environmental quality previously lost may be restored. This directive also provides guidance to Federal, State, and local agencies and the public in commenting on statements prepared under these guidelines.

(b) Pursuant to section 204(3) of the Act the Council on Environmental Quality (hereafter "the Council") is assigned the duty and function of reviewing and appraising the programs and activities of the Federal Government, in the light of the Act's policy, for the purpose of determining the extent to which such programs and activities are contributing to the achievement of such policy, and to make recommendations to the President with respect thereto. Section 102(2)(B) of the Act directs all Federal agencies to identify and develop methods and procedures, in consultation with the Council, to insure that unquantified environmental values be given appropriate consideration in decisionmaking along with economic and technical considerations; section 102(2)(C) of the Act directs that copies of all environmental impact statements be filed with the Council; and section 102(2)(H) directs all Federal agencies to assist the Council in the performance of its functions. These provisions have been supplemented in sections 3(h) and (i) of Executive Order 11514 by directions that the Council issue guidelines to Federal agencies for preparation of environmental impact statements and such other instructions to agencies and requests for reports and information as may be required to carry out the Council's responsibilities under the Act.

§ 1500.2 Policy.

(a) As early as possible and in all cases prior to agency decision concerning recommendations or favorable reports on proposals for (1) legislation significantly affecting the quality of the human environment (see §§ 1500.5(i) and 1500.12) (hereafter "legislative actions") and (2) all other major Federal actions significantly affecting the quality of the human environment (hereafter "administrative actions"), Federal agencies will, in consultation with other appropriate Federal, State and local agencies and the public assess in detail the potential environmental impact.

(b) Initial assessments of the environmental impacts of proposed action should be undertaken concurrently with initial technical and economic studies and, where required, a draft environmental impact statement prepared and circulated for comment in time to accompany the proposal through the existing agency review processes for such action. In this process, Federal agencies shall: (1) Provide for circulation of draft environmental statements to other Federal, State, and local agencies and for their availability to the public in accordance with the provisions of these guidelines; (2) consider the comments of the agencies and the public; and (3) issue final environmental impact statements responsive to the comments received. The purpose of this assessment and consultation process is to provide agencies and other decisionmakers as well as members of the public with an understanding of the potential environmental effects of proposed actions, to avoid or minimize adverse effects wherever possible, and to restore or enhance environmental quality to the fullest extent practicable. In particular, agencies should use the environmental impact statement process to explore alternative actions that will avoid or minimize adverse impacts and to evaluate both the long- and short-range implications of proposed actions to man, his physical and social surroundings, and to nature. Agencies should consider the results of their environmental assessments along with their assessments of the net economic, technical and other benefits of proposed actions and use all practicable means, consistent with other essential considerations of national policy, to restore environmental quality as well as to avoid or minimize undesirable consequences for the environment.

§ 1500.3 Agency and OMB procedures.

(a) Pursuant to section 2(f) of Executive Order 11514, the heads of Federal agencies have been directed to proceed with measures required by section 102(2)(C) of the Act. Previous guidelines of the Council directed each agency to establish its own formal procedures for (1) identifying those agency actions requiring environmental statements, the appropriate time prior to decision for the consultations required by section 102(2)(C) and the agency review process for which environmental statements are to be available, (2) obtaining information required in their preparation, (3) designating the officials who are to be responsible for the statements, (4) consulting with and taking account of the comments of appropriate Federal, State and local agencies and the public, including obtaining the comment of the Administrator of the Environmental Protection Agency when required under section 309 of the Clean Air Act, as amended, and (5) meeting the requirements of section 2(b) of Executive Order 11514 for providing timely public information on Federal plans and programs with environmental impact. Each agency, including both departmental and subdepartmental components having such procedure, shall review its procedures and shall revise them, in consultation with the Council, as may be necessary in order to respond to requirements imposed by these revised guidelines as well as by such previous directives. After such consultation, proposed revisions of such agency procedures shall be published in the FEDERAL REGISTER no later than October 30, 1973. A minimum 45-day period for public comment shall be provided, followed by publication of final procedures no later than forty-five (45) days after the conclusion of the comment period. Each agency shall submit seven (7) copies of all such procedures to the Council. Any future revision of such agency procedures shall similarly be proposed and adopted only after prior consultation with the Council and, in the case of substantial revision, opportunity for public comment. All revisions shall be published in the FEDERAL REGISTER,

(b) Each Federal agency should consult, with the assistance of the Council and the Office of Management and Budget if desired, with other appropriate Federal agencies in the development and revision of the above procedures so as to achieve consistency in dealing with similar activities and to assure effective coordination among agencies in their review of proposed activities. Where applicable, State and local review of such agency procedures should be conducted pursuant to procedures established by Office of Management and Budget Circular No. A-85.

(c) Existing mechanisms for obtaining the views of Federal, State, and local agencies on proposed Federal actions should be utilized to the maximum extent practicable in dealing with environmental matters. The Office of Management and Budget will issue instructions, as necessary, to take full advantage of such existing mechanisms.

§ 1500.4 Federal agencies included; effect of the Act on existing agency mandates.

(a) Section 102(2)(C) of the Act applies to all agencies of the Federal Government. Section 102 of the Act provides that "to the fullest extent possible: (1) The policies, regulations, and public laws of the United States shall be interpreted and administered in accordance with the policies set forth in this Act," and section 105 of the Act provides that "the policies and goals set forth in this Act are supplementary to those set forth in existing authorizations of Federal agencies." This means that each agency shall interpret the provisions of the Act as a supplement to its existing authority and as a mandate to view traditional policies and missions in the light of the Act's national environmental objectives. In accordance with this purpose, agencies should continue to review their policies, procedures, and regulations and to revise them as necessary to ensure full compliance with the purposes and provisions of the Act. The phrase "to the fullest extent possible" in section 102 is meant to make clear that each agency of the Federal Government shall comply with that section unless existing law applicable to the agency's operations expressly prohibits or makes compliance impossible.

§ 1500.5 Types of actions covered by the Act.

(a) "Actions" include but are not limited to:

(1) Recommendations or favorable reports relating to legislation including requests for appropriations. The requirement for following the section 102(2)(C) procedure as elaborated in these guidelines applies to both (i) agency recommendations on their own proposals for legislation (see § 1500.12); and (ii) agency reports on legislation initiated elsewhere. In the latter case only the agency which has primary responsibility for the subject matter involved will prepare an environmental statement.

(2) New and continuing projects and program activities: directly undertaken by Federal agencies; or supported in whole or in part through Federal contracts, grants, subsidies, loans, or other forms of funding assistance (except where such assistance is solely in the form of general revenue sharing funds, distributed under the State and Local Fiscal Assistance Act of 1972, 31 U.S.C. 1221 et. seq. with no Federal agency control over the subsequent use of such funds); or in-

volving a Federal lease, permit, license certificate or other entitlement for use.

(3) The making, modification, or establishment of regulations, rules, procedures, and policy.

§ 1500.6 Identifying major actions significantly affecting the environment.

(a) The statutory clause "major Federal actions significantly affecting the quality of the human environment" is to be construed by agencies with a view to the overall, cumulative impact of the action proposed, related Federal actions and projects in the area, and further actions contemplated. Such actions may be localized in their impact, but if there is potential that the environment may be significantly affected, the statement is to be prepared. Proposed major actions, the environmental impact of which is likely to be highly controversial, should be covered in all cases. In considering what constitutes major action significantly affecting the environment, agencies should bear in mind that the effect of many Federal decisions about a project or complex of projects can be individually limited but cumulatively considerable. This can occur when one or more agencies over a period of years puts into a project individually minor but collectively major resources, when one decision involving a limited amount of money is a precedent for action in much larger cases or represents a decision in principle about a future major course of action, or when several Government agencies individually make decisions about partial aspects of a major action. In all such cases, an environmental statement should be prepared if it is reasonable to anticipate a cumulatively significant impact on the environment from Federal action. The Council, on the basis of a written assessment of the impacts involved, is available to assist agencies in determining whether specific actions require impact statements.

(b) Section 101(b) of the Act indicates the broad range of aspects of the environment to be surveyed in any assessment of significant effect. The Act also indicates that adverse significant effects include those that degrade the quality of the environment, curtail the range of beneficial uses of the environment, and serve short-term, to the disadvantage of long-term, environmental goals. Significant effects can also include actions which may have both beneficial and detrimental effects, even if on balance the agency believes that the effect will be beneficial. Significant effects also include secondary effects as described more fully, for example, in § 1500.8(a)(iii)(B). The significance of a proposed action may also vary with the setting, with the result that an action that would have little impact in an urban area may be significant in a rural setting or vice versa. While a precise definition of environmental "significance," valid in all contexts, is not possible, effects to be considered in assessing significance include, but are not limited to, those outlined in Appendix II of these guidelines.

(c) Each of the provisions of the Act, except section 102(2)(C), applies to all Federal agency actions. Section 102(2)(C) requires the preparation of a detailed environmental impact statement in the case of "major Federal actions significantly affecting the quality of the human environment." The identification of major actions significantly affecting the environment is the responsibility of each Federal agency, to be carried out against the background of its own particular operations. The action must be a (1) "major" action, (2) which is a "Federal action," (3) which has a "significant" effect, and (4) which involves the "quality of the human environment." The words "major" and "significantly" are intended to imply thresholds of importance and impact that must be met before a statement is required. The action causing the impact must also be one where there is sufficient Federal control and responsibility to constitute "Federal action" in contrast to cases where such Federal control and responsibility are not present as, for example, when Federal funds are distributed in the form of general revenue sharing to be used by State and local governments (see § 1500.5(ii)). Finally, the action must be one that significantly affects the quality of the human environment either by directly affecting human beings or by indirectly affecting human beings through adverse effects on the environment. Each agency should review the typical classes of actions that it undertakes and, in consultation with the Council, should develop specific criteria and methods for identifying those actions likely to require environmental statements and those actions likely not to require environmental statements. Normally this will involve:

(i) Making an initial assessment of the environmental impacts typically associated with principal types of agency action.

(ii) Identifying on the basis of this assessment, types of actions which normally do, and types of actions which normally do not, require statements.

(iii) With respect to remaining actions that may require statements depending on the circumstances, and those actions determined under the preceding paragraph (C)(4)(ii) of this section as likely to require statements, identifying: (a) what basic information needs to be gathered; (b) how and when such information is to be assembled and analyzed; and (c) on what bases environmental assessments and decisions to prepare impact statements will be made. Agencies may either include this substantive guidance in the procedures issued pursuant to § 1500.3(a) of these guidelines, or issue such guidance as supplemental instructions to aid relevant agency personnel in implementing the impact statement process. Pursuant to § 1500.14 of these guidelines, agencies shall report to the Council by June 30, 1974, on the progress made in developing such substantive guidance.

(d) (1) Agencies should give careful attention to identifying and defining the purpose and scope of the action which would most appropriately serve as the subject of the statement. In many cases, broad program statements will be required in order to assess the environmental effects of a number of individual actions on a given geographical area (e.g., coal leases), or environmental impacts that are generic or common to a series of agency actions (e.g., maintenance or waste handling practices), or the overall impact of a large-scale program or chain of contemplated projects (e.g. major lengths of highway as opposed to small segments). Subsequent statements on major individual actions will be necessary where such actions have significant environmental impacts not adequately evaluated in the program statement.

(2) Agencies engaging in major technology research and development programs should develop procedures

for periodic evaluation to determine when a program statement is required for such programs. Factors to be considered in making this determination include the magnitude of Federal investment in the program, the likelihood of widespread application of the technology, the degree of environmental impact which would occur if the technology were widely applied, and the extent to which continued investment in the new technology, is likely to restrict future alternatives. Statements must be written late enough in the development process to contain meaningful information, but early enough so that this information can practically serve as an input in the decision-making process. Where it is anticipated that a statement may ultimately be required but that its preparation is still premature, the agency should prepare an evaluation briefly setting forth the reasons for its determination that a statement is not yet necessary. This evaluation should be periodically updated, particularly when significant new information becomes available concerning the potential environmental impact of the program. In any case, a statement must be prepared before research activities have reached a stage of investment or commitment to implementation likely to determine subsequent development or restrict later alternatives. Statements or technology research and development programs should include an analysis not only of alternative forms of the same technology that might reduce any adverse environmental impacts, but also of alternative technologies that would serve the same function as the technology under consideration. Efforts should be made to involve other Federal agencies and interested groups with relevant expertise in the preparation of such statements because the impacts and alternatives to be considered are likely to be less well defined than in other types of statements.

(e) In accordance with the policy of the Act and Executive Order 11514 agencies have a responsibility to develop procedures to insure the fullest practicable provision of timely public information and understanding of Federal plans and programs with environmental impact in order to obtain the views of interested parties. In furtherance of this policy, agency procedures should include an appropriate early notice system for informing the public of the decision to prepare a draft environmental statement on proposed administrative actions (and for soliciting comments that may be helpful in preparing the statement) as soon as is practicable after the decision to prepare the statement is made. In this connection, agencies should: (1) maintain a list of administrative actions for which environmental statements are being prepared; (2) revise the list at regular intervals specified in the agency's procedures developed pursuant to § 1500.3(a) of these guidelines (but not less than quarterly) and transmit each such revision to the Council; and (3) make the list available for public inspection on request. The Council will periodically publish such lists in the FEDERAL REGISTER. If an agency decides that an environmental statement is not necessary for a proposed action (i) which the agency has identified pursuant to § 1500.6(c)(4)(ii) as normally requiring preparation of a statement, (ii) which is similar to actions for which the agency has prepared a significant number of statements, (iii) which the agency has previously announced would be the subject of a statement, or (iv) for which the agency has made a negative determination in response to a request from the Council pursuant to § 1500.11(f), the agency shall prepare a publicly available record briefly setting forth the agency's decision and the reasons for that determination. Lists of such negative determinations, and any evaluations made pursuit to § 1500.6 which conclude that preparation of a statement is not yet timely, shall be prepared and made available in the same manner as provided in this subsection for lists of statements under preparation.

§ 1500.7 Preparing draft environmental statements; public hearings.

(a) Each environmental impact statement shall be prepared and circulated in draft form for comment in accordance with the provisions of these guidelines. The draft statement must fulfill and satisfy to the fullest extent possible at the time the draft is prepared the requirements established for final statements by section 102 (2)(C). (Where an agency has an established practice of declining to favor an alternative until public comments on a proposed action have been received, the draft environmental statement may indicate that two or more alternatives are under consideration.) Comments received shall be carefully evaluated and considered in the decision process. A final statement with substantive comments attached shall then be issued and circulated in accordance with applicable provisions of §§ 1500.10, 1500.11, or 1500.12. It is important that draft environmental statements be prepared and circulated for comment and furnished to the Council as early as possible in the agency review process in order to permit agency decisionmakers and outside reviewers to give meaningful consideration to the environmental issues involved. In particular, agencies should keep in mind that such statements are to serve as the means of assessing the environmental impact of proposed agency actions, rather than as a justification for decisions already made. This means that draft statements on administrative actions should be prepared and circulated for comment prior to the first significant point of decision in the agency review process. For major categories of agency action, this point should be identified in the procedures issued pursuant to § 1500.3(a). For major categories of projects involving an applicant and identified pursuant to § 1500.6 (c)(c)(ii) as normally requiring the preparation of a statement, agencies should include in their procedures provisions limiting actions which an applicant is permitted to take prior to completion and review of the final statement with respect to his application.

(b) Where more than one agency (1) directly sponsors an action, or is directly involved in an action through funding, licenses, or permits, or (2) is involved in a group of actions directly related to each other because of their functional interdependence and geographical proximity, consideration should be given to preparing one statement for all the Federal actions involved (see § 1500.6(d)(1)). Agencies in such cases should consider the possibility of joint preparation of a statement by all agencies concerned, or designation of a single "lead agency" to assume supervisory responsibility for preparation of the statement. Where a lead agency prepares the statement, the other agencies involved should provide assistance with respect to their areas of jurisdiction and expertise. In either case, the statement should contain an environmental assessment of the full range of Federal actions involved, should reflect the views of all participating agencies, and should be prepared before major or irreversible actions have been taken by any of the participating agencies. Factors relevant in determining an appropriate lead agency include the time sequence in which the agencies become involved, the magnitude of their respective involvement, and their relative expertise with respect to the project's environmental effects. As necessary, the Council will assist in resolving questions of responsibility for statement preparation in the case of multi-agency actions. Federal Regional Councils, agencies and the public are encouraged to bring to the attention of the Council and other relevant agencies appropriate situations where a geographic or regionally focused statement would be desirable because of the cumulative environmental effects likely to result from multi-agency actions in the area.

(c) Where an agency relies on an applicant to submit initial environmental information, the agency should assist the applicant by outlining the types of information required. In all cases, the agency should make its own evaluation of the environmental issues and take responsibility for the scope and content of draft and final environmental statements.

(d) Agency procedures developed pursuant to § 1500.3(a) of these guidelines should indicate as explicitly as possible those types of agency decisions or actions which utilize hearings as part of the normal agency review process, either as a result of statutory requirement or agency practice. To the fullest extent possible, all such hearings shall include consideration of the environmental aspects of the proposed action. Agency procedures shall also specifically include provision for public hearings on major actions with environmental impact, whenever appropriate, and for providing the public with relevant information, including information on alternative courses of action. In deciding whether a public hearing is appropriate, an agency should consider: (1) The magnitude of the proposal in terms of economic costs, the geographic area involved, and the uniqueness or size of commitment of the resources involved; (2) the degree of interest in the proposal, as evidenced by requests from the public and from Federal, State and local authorities that a hearing be held; (3) the complexity of the issue and the likelihood that information will be presented at the hearing which will be of assistance to the agency in fulfilling its responsibilities under the Act; and (4) the extent to which public involvement already has been achieved through other means, such as earlier public hearings, meetings with citizen representatives, and/or written comments on the proposed action. Agencies should make any draft environmental statements to be issued available to the public at least fifteen (15) days prior to the time of such hearings.

§ 1500.8 Content of environmental statements.

(a) The following points are to be covered:

(1) A description of the proposed

action, a statement of its purposes, and a description of the environment affected, including information, summary technical data, and maps and diagrams where relevant, adequate to permit an assessment of potential environmental impact by commenting agencies and the public. Highly technical and specialized analyses and data should be avoided in the body of the draft impact statement. Such materials should be attached as appendices or footnoted with adequate bibliographic references. The statement should also succinctly describe the environment of the area affected as it exists prior to a proposed action, including other Federal activities in the area affected by the proposed action which are related to the proposed action. The interrelationships and cumulative environmental impacts of the proposed action and other related Federal projects shall be presented in the statement. The amount of detail provided in such descriptions should be commensurate with the extent and expected impact of the action, and with the amount of information required at the particular level of decisionmaking (planning, feasibility, design, etc.). In order to ensure accurate descriptions and environmental assessments, site visits should be made where feasible. Agencies should also take care to identify, as appropriate, population and growth characteristics of the affected area and any population and growth assumptions used to justify the project or program or to determine secondary population and growth impacts resulting from the proposed action and its alternatives (see paragraph (a)(1)(3)(ii), of this section). In discussing these population aspects, agencies should give consideration to using the rates of growth in the region of the project contained in the projection compiled for the Water Resources Council by the Bureau of Economic Analysis of the Department of Commerce and the Economic Research Service of the Department of Agriculture (the "OBERS" projection). In any event it is essential that the sources of data used to identify, quantify or evaluate any and all environmental consequences be expressly noted.

(2) The relationship of the proposed action to land use plans, policies, and controls for the affected area. This requires a discussion of how the proposed action may conform or conflict with the objectives and specific terms of approved or proposed Federal, State, and local land use plans, policies, and controls, if any, for the area affected including those developed in response to the Clean Air Act or the Federal Water Pollution Control Act Amendments of 1972. Where a conflict or inconsistency exists, the statement should describe the extent to which the agency has reconciled its proposed action with the plan, policy or control, and the reasons why the agency has decided to proceed notwithstanding the absence of full reconciliation.

(3) The probable impact of the proposed action on the environment.

(i) This requires agencies to assess the positive and negative effects of the proposed action as it affects both the national and international environment. The attention given to different environmental factors will vary according to the nature, scale, and location of proposed actions. Among factors to consider should be the potential effect of the action on such aspects of the environment as those listed in Appendix II of these guidelines. Primary attention should be given in the statement to discussing those factors most evidently impacted by the proposed action.

(ii) Secondary or indirect, as well as primary or direct, consequences for the environment should be included in the analysis. Many major Federal actions, in particular those that involve the construction or licensing of infrastructure investments (e.g., highways, airports, sewer systems, water resource projects, etc.), stimulate or induce secondary effects in the form of associated investments and changed patterns of social and economic activities. Such secondary effects, through their impacts on existing community facilities and activities, through inducing new facilities and activities, or through changes in natural conditions, may often be even more substantial than the primary effects of the original action itself. For example, the effects of the proposed action on population and growth may be among the more significant secondary effects. Such population and growth impacts should be estimated if expected to be significant (using data identified as indicated in § 1500.8(a)(1) and an assessment made of the effect of any possible change in population patterns or growth upon the resource base, including land use, water, and public services, of the area in question.

(4) *Alternatives to the proposed action, including, where relevant, those not within the existing authority of the responsible agency.* (Section 102(2)(D) of the Act requires the responsible agency to "study, develop, and describe appropriate alternatives to recommended courses of action in any proposal which involves unresolved conflicts concerning alternative uses of available resources"). A rigorous exploration and objective evaluation of the environmental impacts of all reasonable alternative actions, particularly those that might enhance environmental quality or avoid some or all of the adverse environmental effects, is essential. Sufficient analysis of such alternatives and their environmental benefits, costs and risks should accompany the proposed action through the agency review process in order not to foreclose prematurely options which might enhance environmental quality or have less detrimental effects. Examples of such alternatives include: the alternative of taking no action or of postponing action pending further study; alternatives requiring actions of a significantly different nature which would provide similar benefits with different environmental impacts (e.g., nonstructural alternatives to flood control programs, or mass transit alternatives to highway construction); alternatives related to different designs or details of the proposed action which would prevent different environmental impacts (e.g., cooling ponds vs. cooling towers for a power plant or alternatives that will significantly conserve energy); alternative measures to provide for compensation of fish and wildlife losses, including the acquisition of land, waters, and interests therein. In each case, the analysis should be sufficiently detailed to reveal the agency's comparative evaluation of the environmental benefits, costs and risks of the proposed action and each reasonable alternative. Where an existing impact statement already contains such an analysis, its treatment of alternatives may be incorporated provided that such treatment is current and relevant to the precise purpose of the proposed action.

(5) *Any probable adverse environ-*

mental effects which cannot be avoided (such as water or air pollution, undesirable land use patterns, damage to life systems, urban congestion, threats to health or other consequences adverse to the environmental goals set out in section 101(b) of the Act). This should be a brief section summarizing in one place those effects discussed in paragraph (a)(3) of this section that are adverse and unavoidable under the proposed action. Included for purposes of contrast should be a clear statement of how other avoidable adverse effects discussed in paragraph (a)(2) of this section will be mitigated.

(6) The relationship between local short-term uses of man's environment and the maintenance and enhancement of long-term productivity. This section should contain a brief discussion of the extent to which the proposed action involves tradeoffs between short-term environmental gains at the expense of long-term losses, or vice versa, and a discussion of the extent to which the proposed action forecloses future options. In this context short-term and long-term do not refer to any fixed time periods, but should be viewed in terms of the environmentally significant consequences of the proposed action.

(7) Any irreversible and irretrievable commitments of resources that would be involved in the proposed action should it be implemented. This requires the agency to identify from its survey of unavoidable impacts in paragraph (a)(5) of this section the extent to which the action irreversibly curtails the range of potential uses of the environment. Agencies should avoid construing the term "resources" to mean only the labor and materials devoted to an action. "Resources" also means the natural and cultural resources committed to loss or destruction by the action.

(8) An indication of what other interests and considerations of Federal policy are thought to offset the adverse environmental effects of the proposed action identified pursuant to paragraphs (a)(3) and (5) of this section. The statement should also indicate the extent to which these stated countervailing benefits could be realized by following reasonable alternatives to the proposed action (as identified in paragraph (a)(4) of this section) that would avoid some or all of the adverse environmental effects. In this connection, agencies that prepare cost-benefit analyses of proposed actions should attach such analyses, or summaries thereof, to the environmental impact statement, and should clearly indicate the extent to which environmental costs have not been reflected in such analyses.

(b) In developing the above points agencies should make every effort to convey the required information succinctly in a form easily understood, both by members of the public and by public decisionmakers, giving attention to the substance of the information conveyed rather than to the particular form, or length, or detail of the statement. Each of the above points, for example, need not always occupy a distinct section of the statement if it is otherwise adequately covered in discussing the impact of the proposed action and its alternatives— which items should normally be the focus of the statement. Draft statements should indicate at appropriate points in the text any underlying studies, reports, and other information obtained and considered by the agency in preparing the statement including any cost-benefit analyses prepared by the agency, and reports of consulting agencies under the Fish and Wildlife Coordination Act, 16 U.S.C. 661 et seq,. and the National Historic Preservation Act of 1966, 16 U.S.C. 470 et seq., where such consultation has taken place. In the case of documents not likely to be easily accessible (such as internal studies or reports), the agency should indicate how such information may be obtained. If such information is attached to the statement, care should be taken to ensure that the statement remains an essentially self-contained instrument, capable of being understood by the reader without the need for undue cross reference.

(c) Each environmental statement should be prepared in accordance with the precept in section 102(2)(A) of the Act that all agencies of the Federal Government "utilize a systematic, interdisciplinary approach which will insure the integrated use of the natural and social sciences and the environmental design arts in planning and decisionmaking which may have an impact on man's environment." Agencies should attempt to have relevant disciplines represented on their own staffs; where this is not feasible they should make appropriate use of relevant Federal, State, and local agencies or the professional services of universities and outside consultants. The interdisciplinary approach should not be limited to the preparation of the environmental impact statement, but should also be used in the early planning stages of the proposed action. Early application of such an approach should help assure a systematic evaluation of reasonable alternative courses of action and their potential social, economic, and environmental consequences.

(d) Appendix I prescribes the form of the summary sheet which should accompany each draft and final environmental statement.

§ 1500.9 **Review of draft environmental statements by Federal, Federal-State, State, and local agencies and by the public.**

(a) *Federal agency review.* (1) *In general.* A Federal agency considering an action requiring an environmental statement should consult with, and (on the basis of a draft environmental statement for which the agency takes responsibility) obtain the comment on the environmental impact of the action of Federal and Federal-State agencies with jurisdiction by law or special expertise with respect to any environmental impact involved. These Federal and Federal-State agencies and their relevant areas of expertise include those identified in Appendices II and III to these guidelines. It is recommended that the listed departments and agencies establish contact points, which may be regional offices, for providing comments on the environmental statements. The requirement in section 102(2)(C) to obtain comment from Federal agencies having jurisdiction or special expertise is in addition to any specific statutory obligation of any Federal agency to coordinate or consult with any other Federal or State agency. Agencies should, for example, be alert to consultation requirements of the Fish and Wildlife Coordination Act, 16 U.S.C. 661 et seq., and the National Historic Preservation Act of 1966, 16 U.S.C. 470 et seq. To the extent possible, statements or findings concerning environmental impact required by other statutes, such as section 4(f) of the Department of Transportation Act of 1966, 49 U.S.C. 1653(f), or section 106 of the National Historic Preservation Act of 1966, should be combined with compliance with the environmental impact statement requirements of section 102(2)(C) of the Act to yield a single document which meets all applicable requirements. The Advisory Council on Historic Preservation, the Department of Transportation, and the Department of the Interior, in consultation with the Council, will issue any necessary supplementing instructions for furnishing information or findings not forthcoming under the environmental impact statement process.

(b) *EPA review.* Section 309 of the Clean Air Act, as amended (42 U.S.C. § 1857h–7), provides that the Administrator of the Environmental Protection Agency shall comment in writing on the environmental impact of any matter relating to his duties and responsibilities, and shall refer to the Council any matter that the Administrator determines is unsatisfactory from the standpoint of public health or welfare or environmental quality. Accordingly, wherever an agency action related to air or water quality, noise abatement and control, pesticide regulation, solid waste disposal, generally applicable environmental radiation criteria and standards, or other provision of the authority of the Administrator is involved, Federal agencies are required to submit such proposed actions and their environmental impact statements, if such have been prepared, to the Administrator for review and comment in writing. In all cases where EPA determines that proposed agency action is environmentally unsatisfactory, or where EPA determines that an environmental statement is so inadequate that such a determination cannot be made, EPA shall publish its determination and notify the Council as soon as practicable. The Administrator's comments shall constitute his comments for the purposes of both section 309 of the Clean Air Act and section 102(2)(C) of the National Environmental Policy Act.

(c) State and local review. Office of Management and Budget Circular No. A–95 (Revised) through its system of State and areawide clearinghouses provides a means for securing the views of State and local environmental agencies, which can assist in the preparation and review of environmental impact statements. Current instructions for obtaining the views of such agencies are contained in the joint OMB–CEQ memorandum attached to these guidelines as Appendix IV. A current listing of clearinghouses is issued periodically by the Office of Management and Budget.

(d) *Public review.* The procedures established by these guidelines are designed to encourage public participation in the impact statement process at the earliest possible time. Agency procedures should make provision for facilitating the comment of public and private organizations and individuals by announcing the availability of draft environmental statements and by making copies available to organizations and individuals that request an opportunity to comment. Agencies should devise methods for publicizing the existence of draft statements, for example, by publication of notices in local newspapers or by maintaining a list of groups, including relevant conservation commissions, known to be interested in the agency's activities and directly notifying such groups of the existence of a draft statement, or sending them a copy, as soon as it has been prepared. A copy of the draft statement should in all cases be sent to any applicant whose project is the subject of the statement. Materials to be made available to the public shall be provided without charge to the extent practicable, or at a fee which is not more than the actual cost of reproducing copies required to be sent to other Federal agencies, including the Council.

(e) *Responsibilities of commenting entities.* (1) Agencies and members of the public submitting comments on proposed actions on the basis of draft environmental statements should endeavor to make their comments as specific, substantive, and factual as possible without undue attention to matters of form in the impact statement. Although the comments need not conform to any particular format, it would assist agencies reviewing comments if the comments were organized in a manner consistent with the structure of the draft statement. Emphasis should be placed on the assessment of the environmental impacts of the proposed action, and the acceptability of those impacts on the quality of the environment, particularly as contrasted with the impacts of reasonable alternatives to the action. Commenting entities may recommend modifications to the proposed action and/or new alternatives that will enhance environmental quality and avoid or minimize adverse environmental impacts.

(2) Commenting agencies should indicate whether any of their projects not identified in the draft statement are sufficiently advanced in planning and related environmentally to the proposed action so that a discussion of the environmental interrelationships should be included in the final statement (see § 1500.8(a)(1)). The Council is available to assist agencies in making such determinations.

(3) Agencies and members of the public should indicate in their comments the nature of any monitoring of the environmental effects of the proposed project that appears particularly appropriate. Such monitoring may be necessary during the construction, startup, or operation phases of the project. Agencies with special expertise with respect to the environmental impacts involved are encouraged to assist the sponsoring agency in the establishment and operation of appropriate environmental monitoring.

(f) Agencies seeking comment shall establish time limits of not less than forty-five (45) days for reply, after which it may be presumed, unless the agency or party consulted requests a specified extension of time, that the agency or party consulted has no comment to make. Agencies seeking comment should endeavor to comply with requests for extensions of time of up to fifteen (15) days. In determining an appropriate period for comment, agencies should consider the magnitude and complexity of the statement and the extent of citizen interest in the proposed action.

§ 1500.10 **Preparation and circulation of final environmental statements.**

(a) Agencies should make every effort to discover and discuss all major points of view on the environmental effects of the proposed action and its alternatives in the draft statement itself. However, where opposing professional views and responsible opinion have been overlooked in the draft statement and are brought to the agency's attention through the commenting process, the agency should review the environmental effects of the action in light of those views and should make a meaningful reference in the final statement to the existence of any responsible opposing view not adequately discussed in the draft statement, indicating the agency's response to the issues raised. All substantive comments received on the draft (or summaries thereof where response has been exceptionally voluminous) should be attached to the final statement, whether or not each such comment is thought to merit individual discussion by the agency in the text of the statement.

(b) Copies of final statements, with comments attached, shall be sent to all Federal, State, and local agencies and private organizations that made substantive comments on the draft statement and to individuals who requested a copy of the final statement, as well as any applicant whose project is the subject of the statement. Copies of final statements shall in all cases be sent to the Environmental Protection Agency to assist it in carrying out its responsibilities under section 309 of the Clean Air Act. Where the number of comments on a draft statement is such that distribution of the final statement to all commenting entities appears impracticable, the agency shall consult with the Council concerning alternative arrangements for distribution of the statement.

§ 1500.11 Transmittal of statements to the Council; minimum periods for review; requests by the Council.

(a) As soon as they have been prepared, ten (10) copies of draft environmental statements, five (5) copies of all comments made thereon (to be forwarded to the Council by the entity making comment at the time comment is forwarded to the responsible agency), and ten (10) copies of the final text of environmental statements (together with the substance of all comments received by the responsible agency from Federal, State, and local agencies and from private organizations and individuals) shall be supplied to the Council. This will serve to meet the statutory requirement to make environmental statements available to the President. At the same time that copies of draft and final statements are sent to the Council, copies should also be sent to relevant commenting entities as set forth in §§ 1500.9 and 1500.10(b) of these guidelines.

(b) To the maximum extent practicable no administrative action subject to section 102(2)(C) is to be taken sooner than ninety (90) days days after a draft environmental statement has been circulated for comment, furnished to the Council and, except where advance public disclosure will result in significantly increased costs of procurement to the Government, made available to the public pursuant to these guidelines; neither should such administrative action be taken sooner than thirty (30) days after the final text of an environmental statement (together with comments) has been made available to the Council, commenting agencies, and the public. In all cases, agencies should allot a sufficient review period for the final statement so as to comply with the statutory requirement that the "statement and the comments and views of appropriate Federal, State, and local agencies * * * accompany the proposal through the existing agency review processes." If the final text of an environmental statement is filed within ninety (90) days after a draft statement has been circulated for comment, furnished to the Council and made public pursuant to this section of these guidelines, the minimum thirty (30) day period and the ninety (90) day period may run concurrently to the extent that they overlap. An agency may at any time supplement or amend a draft or final environmental statement, particularly when substantial changes are made in the proposed action, or significant new information becomes available concerning its environmental aspects. In such cases, the agency should consult with the Council with respect to the possible need for or desirability of recirculation of the statement for the appropriate period.

(c) The Council will publish weekly in the FEDERAL REGISTER lists of environmental statements received during the preceding week that are available for public comment. The date of publication of such lists shall be the date from which the minimum periods for review and advance availability of statements shall be calculated.

(d) The Council's publication of notice of the availability of statements is in addition to the agency's responsibility, as described in § 1500.9(d) of these guidelines, to insure the fullest practicable provision of timely public information concerning the existence and availability of environmental statements. The agency responsible for the environmental statement is also responsible for making the statement, the comments received, and any underlying documents available to the public pursuant to the provisions of the Freedom of Information Act (5 U.S.C., 552), without regard to the exclusion of intra- or interagency memoranda when such memoranda transmit comments of Federal agencies on the environmental impact of the proposed action pursuant to § 1500.9 of these guidelines. Agency procedures prepared pursuant to § 1500.3(a) of these guidelines shall implement these public information requirements and shall include arrangements for availability of environmental statements and comments at the head and appropriate regional offices of the responsible agency and at appropriate State and areawide clearinghouses unless the Governor of the State involved designates to the Council some other point for receipt of this information. Notice of such designation of an alternate point for receipt of this information will be included in the Office of Management and Budget listing of clearinghouses referred to in § 1500.9(c).

(e) Where emergency circumstances make it necessary to take an action with significant environmental impact without observing the provisions of these guidelines concerning minimum periods for agency review and advance availability of environmental statements, the Federal agency proposing to take the action should consult with the Council about alternative arrangements. Similarly where there are overriding considerations of expense to the Government or impaired program effectiveness, the responsible agency should consult with the Council concerning appropriate modifications of the minimum periods.

(f) In order to assist the Council in fulfilling its responsibilities under the Act and under Executive Order 11514, all agencies shall (as required by section 102(2)(H) of the Act and section 3(i) of Executive Order 11514) be responsive to requests by the Council for reports and other information dealing with issues arising in connection with the implementation of the Act. In particular, agencies shall be responsive to a request by the Council for the preparation and circulation of an environmental statement, unless the agency determines that such a statement is not required, in which case the agency shall prepare an environmental assessment and a publicly available record briefly setting forth the reasons for its determination. In no case, however, shall the Council's silence or failure to comment or request preparation, modification, or recirculation of an environmental statement or to take other action with respect to an environmental statement be construed as bearing in any way on the question of the legal requirement for or the adequacy of such statement under the Act.

§ 1500.12 Legislative actions.

(a) The Council and the Office of Management and Budget will cooperate in giving guidance as needed to assist agencies in identifying legislative items believed to have environmental significance. Agencies should prepare impact statements prior to submission of their legislative proposals to the Office of Management and Budget. In this regard, agencies should identify types of repetitive legislation requiring environmental impact statements (such as certain types of bills affecting transportation policy or annual construction authorizations).

(b) With respect to recommendations or reports on proposals for legislation to which section 102(2)(C) applies, the final text of the environmental statement and comments thereon should be available to the Congress and to the public for consideration in connection with the proposed legislation or report. In cases where the scheduling of congressional hearings on recommendations or reports on proposals for legislation which the Federal agency has forwarded to the Congress does not allow adequate time for the completion of a final text of an environmental statement (together with comments), a draft environmental statement may be furnished to the Congress and made available to the public pending transmittal of the comments as received and the final text.

§ 1500.13 Application of section 102 (2)(C) procedure to existing projects and programs.

Agencies have an obligation to reassess ongoing projects and programs in order to avoid or minimize adverse environmental effects. The section 102(2)(C) procedure shall be applied to further major Federal actions having a significant effect on the environment even though they arise from projects or programs initiated prior to enactment of the Act on January 1, 1970. While the status of the work and degree of completion may be considered in determining whether to proceed with the project, it is essential that the environmental impacts of proceeding are reassessed pursuant to the Act's policies and procedures and, if the project or program is continued, that further incremental major actions be shaped so as to enhance and restore environmental quality as well as to avoid or minimize adverse environmental consequences. It is also important in further action that account be taken of environmental consequences not fully evaluated at the outset of the project or program.

§ 1500.14 Supplementary guidelines; evaluation of procedures.

(a) The Council after examining environmental statements and agency procedures with respect to such statements will issue such supplements to these guidelines as are necessary.

(b) Agencies will continue to assess their experience in the implementation of the section 102(2)(C) provisions of the Act and in conforming with these guidelines and report thereon to the Council by June 30, 1974. Such reports should include an identification of the problem areas and suggestions for revision or clarification of these guidelines to achieve effective coordination of views on environmental aspects (and alternatives, where appropriate) of proposed actions without imposing unproductive administrative procedures. Such re-reports shall also indicate what progress the agency has made in developing substantive criteria and guidance for making environmental assessments as required by § 1500.6(c) of this directive and by section 102(2)(B) of the Act.

Effective date. The revisions of these guidelines shall apply to all draft and final impact statements filed with the Council after January 28, 1973.

RUSSELL E. TRAIN,
Chairman

APPENDIX I—SUMMARY TO ACCOMPANY DRAFT AND FINAL STATEMENTS

(Check one) ( ) Draft. ( ) Final Environmental Statement.

Name of responsible Federal agency (with name of operating division where appropriate). Name, address, and telephone number of individual at the agency who can be contacted for additional information about the proposed action or the statement.

1. Name of action (Check one) ( ) Administrative Action. ( ) Legislative Action.

2. Brief description of action and its purpose. Indicate what States (and counties) particularly affected, and what other proposed Federal actions in the area, if any, are discussed in the statement.

3. Summary of environmental impacts and adverse environmental effects.

4. Summary of major alternatives considered.

5. (For draft statements) List all Federal, State, and local agencies and other parties from which comments have been requested. (For final statements) List all Federal, State, and local agencies and other parties from which written comments have been received.

6. Date draft statement (and final environmental statement, if one has been issued) made available to the Council and the public.

APPENDIX II—AREAS OF ENVIRONMENTAL IMPACT AND FEDERAL AGENCIES AND FEDERAL STATE AGENCIES [1] WITH JURISDICTION BY LAW OR SPECIAL EXPERTISE TO COMMENT THEREON [2]

AIR

Air Quality

Department of Agriculture—
　Forest Service (effects on vegetation)
Atomic Energy Commission (radioactive substances)
Department of Health, Education, and Welfare
Environmental Protection Agency
Department of the Interior—
　Bureau of Mines (fossil and gaseous fuel combustion)

[1] River Basin Commissions (Delaware, Great Lakes, Missouri, New England, Ohio, Pacific Northwest, Souris-Red-Rainy, Susquehanna, Upper Mississippi) and similar Federal-State agencies should be consulted on actions affecting the environment of their specific geographic jurisdictions.

[2] In all cases where a proposed action will have significant international environmental effects, the Department of State should be consulted, and should be sent a copy of any draft and final impact statement which covers such action.

Bureau of Sport Fisheries and Wildlife (effect on wildlife)
Bureau of Outdoor Recreation (effects on recreation)
Bureau of Land Management (public lands)
Bureau of Indian Affairs (Indian lands)
National Aeronautics and Space Administration (remote sensing, aircraft emissions)
Department of Transportation—
Assistant Secretary for Systems Development and Technology (auto emissions)
Coast Guard (vessel emissions)
Federal Aviation Administration (aircraft emissions)

Weather Modification

Department of Agriculture—
Forest Service
Department of Commerce—
National Oceanic and Atmospheric Administration
Department of Defense—
Department of the Air Force
Department of the Interior—
Bureau of Reclamation

WATER RESOURCES COUNCIL

WATER

Water Quality

Department of Agriculture—
Soil Conservation Service
Forest Service
Atomic Energy Commission (radioactive substances)
Department of the Interior—
Bureau of Reclamation
Bureau of Land Management (public lands)
Bureau of Indian Affairs (Indian lands)
Bureau of Sport Fisheries and Wildlife
Bureau of Outdoor Recreation
Geological Survey
Office of Saline Water
Environmental Protection Agency
Department of Health, Education, and Welfare
Department of Defense—
Army Corps of Engineers
Department of the Navy (ship pollution control)

National Aeronautics and Space Administration (remote sensing)
Department of Transportation—
Coast Guard (oil spills, ship sanitation)
Department of Commerce—
National Oceanic and Atmospheric Administration
Water Resources Council
River Basin Commissions (as geographically appropriate)

Marine Pollution, Commercial Fishery Conservation, and Shellfish Sanitation

Department of Commerce—
National Oceanic and Atmospheric Administration
Department of Defense
Army Corps of Engineers
Office of the Oceanographer of the Navy
Department of Health, Education, and Welfare
Department of the Interior—
Bureau of Sport Fisheries and Wildlife
Bureau of Outdoor Recreation
Bureau of Land Management (outer continental shelf)
Geological Survey (outer continental shelf)
Department of Transportation—
Coast Guard
Environmental Protection Agency
National Aeronautics and Space Administration (remote sensing)
Water Resources Council
River Basin Commissions (as geographically appropriate)

Waterway Regulation and Stream Modification

Department of Agriculture—
Soil Conservation Service
Department of Defense—
Army Corps of Engineers
Department of the Interior—
Bureau of Reclamation
Bureau of Sport Fisheries and Wildlife
Bureau of Outdoor Recreation
Geological Survey
Department of Transportation—
Coast Guard
Environmental Protection Agency
National Aeronautics and Space Administration (remote sensing)
Water Resources Council

Council on Environmental Quality Guidelines

River Basin Commissions (as geographically appropriate)

FISH AND WILDLIFE

Department of Agriculture—
Forest Service
Soil Conservation Service
Department of Commerce—
National Oceanic and Atmospheric Administration (marine species)
Department of the Interior—
Bureau of Sport Fisheries and Wildlife
Bureau of Land Management
Bureau of Outdoor Recreation
Environmental Protection Agency

SOLID WASTE

Atomic Energy Commission (radioactive waste)
Department of Defense—
Army Corps of Engineers
Department of Health, Education, and Welfare
Department of the Interior—
Bureau of Mines (mineral waste, mine acid waste, municipal solid waste, recycling)
Bureau of Land Management (public lands)
Bureau of Indian Affairs (Indian lands)
Geological Survey (geologic and hydrologic effects)
Office of Saline Water (demineralization)
Department of Transportation—
Coast Guard (ship sanitation)
Environmental Protection Agency
River Basin Commissions (as geographically appropriate)
Water Resources Council

NOISE

Department of Commerce—
National Bureau of Standards
Department of Health, Education, and Welfare
Department of Housing and Urban Development (land use and building materials aspects)
Department of Labor—
Occupational Safety and Health Administration
Department of Transportation—
Assistant Secretary for Systems Development and Technology
Federal Aviation Administration, Office of Noise Abatement
Environmental Protection Agency
National Aeronautics and Space Administration

RADIATION

Atomic Energy Commission
Department of Commerce—
National Bureau of Standards
Department of Health, Education, and Welfare
Department of the Interior
Bureau of Mines (uranium mines)
Mining Enforcement and Safety Administration (uranium mines)
Environmental Protection Agency

HAZARDOUS SUBSTANCES

Toxic Materials

Atomic Energy Commission (radioactive substances)
Department of Agriculture—
Agricultural Research Service
Consumer and Marketing Service
Department of Commerce—
National Oceanic and Atmospheric Administration
Department of Defense
Department of Health, Education, and Welfare
Environmental Protection Agency

Food Additives and Contamination of Foodstuffs

Department of Agriculture—
Consumer and Marketing Service (meat and poultry products)
Department of Health, Education, and Welfare
Environmental Protection Agency

Pesticides

Department of Agriculture—
Agricultural Research Service (biological controls, food and fiber production)
Consumer and Marketing Service
Forest Service
Department of Commerce—
National Oceanic and Atmospheric Administration
Department of Health, Education, and Welfare

Department of the Interior—
Bureau of Sport Fisheries and Wildlife (fish and wildlife effects)
Bureau of Land Management (public lands)
Bureau of Indian Affairs (Indian lands)
Bureau of Reclamation (irrigated lands)
Environmental Protection Agency

Transportation and Handling of Hazardous Materials

Atomic Energy Commission (radioactive substances)
Department of Commerce—
Maritime Administration
National Oceanic and Atmospheric Administration (effects on marine life and the coastal zone)
Department of Defense—
Armed Services Explosive Safety Board
Army Corps of Engineers (navigable waterways)
Department of Transportation—
Federal Highway Administration, Bureau of Motor Carrier Safety
Coast Guard
Federal Railroad Administration
Federal Aviation Administration
Assistant Secretary for Systems Development and Technology
Office of Hazardous Materials
Office of Pipeline Safety
Environmental Protection Agency

ENERGY SUPPLY AND NATURAL RESOURCES DEVELOPMENT

Electric Energy Development, Generation, and Transmission, and Use

Atomic Energy Commission (nuclear)
Department of Agriculture—
Rural Electrification Administration (rural areas)
Department of Defense—
Army Corps of Engineers (hydro)
Department of Health, Education, and Welfare (radiation effects)
Department of Housing and Urban Development (urban areas)
Department of the Interior—
Bureau of Indian Affairs (Indian lands)
Bureau of Land Management (public lands)
Bureau of Reclamation
Power Marketing Administrations

Geological Survey
Bureau of Sport Fisheries and Wildlife
Bureau of Outdoor Recreation
National Park Service
Environmental Protection Agency
Federal Power Commission (hydro, transmission, and supply)
River Basin Commissions (as geographically appropriate)
Tennessee Valley Authority
Water Resources Council

Petroleum Development, Extraction, Refining, Transport, and Use

Department of the Interior—
Office of Oil and Gas
Bureau of Mines
Geological Survey
Bureau of Land Management (public lands and outer continental shelf)
Bureau of Indian Affairs (Indian lands)
Bureau of Sport Fisheries and Wildlife (effects on fish and wildlife)
Bureau of Outdoor Recreation
National Park Service
Department of Transportation (Transport and Pipeline Safety)
Environmental Protection Agency
Interstate Commerce Commission

Natural Gas Development, Production, Transmission, and Use

Department of Housing and Urban Development (urban areas)
Department of the Interior—
Office of Oil and Gas
Geological Survey
Bureau of Mines
Bureau of Land Management (public lands)
Bureau of Indian Affairs (Indian lands)
Bureau of Sport Fisheries and Wildlife
Bureau of Outdoor Recreation
National Park Service
Department of Transportation (transport and safety)
Environmental Protection Agency
Federal Power Commission (production, transmission, and supply)
Interstate Commerce Commission

Coal and Minerals Development, Mining Conversion, Processing, Transport, and Use

Appalachian Regional Commission

Department of Agriculture—
  Forest Service
Department of Commerce
Department of the Interior—
  Office of Coal Research
  Mining Enforcement and Safety Administration
  Bureau of Mines
  Geological Survey
  Bureau of Indian Affairs (Indian lands)
  Bureau of Land Management (public lands)
  Bureau of Sport Fisheries and Wildlife
  Bureau of Outdoor Recreation
  National Park Service
Department of Labor—
  Occupational Safety and Health Administration
Department of Transportation
Environmental Protection Agency
Interstate Commerce Commission
Tennessee Valley Authority

Renewable Resource Development, Production, Management, Harvest, Transport, and Use

Department of Agriculture—
  Forest Service
  Soil Conservation Service
Department of Commerce
Department of Housing and Urban Development (building materials)
Department of the Interior—
  Geological Survey
  Bureau of Land Management (public lands)
  Bureau of Indian Affairs (Indian lands)
  Bureau of Sport Fisheries and Wildlife
  Bureau of Outdoor Recreation
  National Park Service
Department of Transportation
Environmental Protection Agency
Interstate Commerce Commission (freight rates)

Energy and Natural Resources

Department of Agriculture—
  Forest Service
  Soil Conservation Service
Department of Commerce—
  National Bureau of Standards (energy efficiency)
Department of Housing and Urban Development—
  Federal Housing Administration (housing standards)
Department of the Interior—
  Office of Energy Conservation
  Bureau of Mines
  Bureau of Reclamation
  Geological Survey
  Power Marketing Administration
Department of Transportation
Environmental Protection Agency
Federal Power Commission
General Services Administration (design and operation of buildings)
Tennessee Valley Authority

LAND USE AND MANAGEMENT

Land Use Changes, Planning and Regulation of Land Development

Department of Agriculture—
  Forest Service (forest lands)
  Agricultural Research Service (agricultural lands)
Department of Housing and Urban Development
Department of the Interior—
  Office of Land Use and Water Planning
  Bureau of Land Management (public lands)
  Bureau of Indian Affairs (Indian lands)
  Bureau of Sport Fisheries and Wildlife (wildlife refuges)
  Bureau of Outdoor Recreation (recreation lands)
  National Park Service (NPS units)
Department of Transportation
Environmental Protection Agency (pollution effects)
National Aeronautics and Space Administration (remote sensing)
River Basin Commissions (as geographically appropriate)

Public Land Management

Department of Agriculture—
  Forest Service (forests)
Department of Defense
Department of the Interior—
  Bureau of Land Management
  Bureau of Indian Affairs (Indian lands)
  Bureau of Sport Fisheries and Wildlife (wildlife refuges)
  Bureau of Outdoor Recreation (recreation lands)
  National Park Service (NPS units)
Federal Power Commission (project lands)
General Services Administration

National Aeronautics and Space Administration (remote sensing)
Tennessee Valley Authority (project lands)

PROTECTION OF ENVIRONMENTALLY CRITICAL AREAS—FLOODPLAINS, WETLANDS, BEACHES AND DUNES, UNSTABLE SOILS, STEEP SLOPES, AQUIFER RECHARGE AREAS, ETC.

Department of Agriculture—
Agricultural Stabilization and Conservation Service
Soil Conservation Service
Forest Service
Department of Commerce—
National Oceanic and Atmospheric Administration (coastal areas)
Department of Defense—
Army Corps of Engineers
Department of Housing and Urban Development (urban and floodplain areas)
Department of the Interior—
Office of Land Use and Water Planning
Bureau of Outdoor Recreation
Bureau of Reclamation
Bureau of Sport Fisheries and Wildlife
Bureau of Land Management
Geological Survey
Environmental Protection Agency (pollution effects)
National Aeronautics and Space Administration (remote sensing)
River Basins Commissions (as geographically appropriate)
Water Resources Council

LAND USE IN COASTAL AREAS

Department of Agriculture—
Forest Service
Soil Conservation Service (soil stability, hydrology)
Department of Commerce—
National Oceanic and Atmospheric Administration (impact on marine life and coastal zone management)
Department of Defense—
Army Corps of Engineers (beaches, dredge and fill permits, Refuse Act permits)
Department of Housing and Urban Development (urban areas)
Department of the Interior—
Office of Land Use and Water Planning
Bureau of Sport Fisheries and Wildlife
National Park Service
Geological Survey
Bureau of Outdoor Recreation
Bureau of Land Management (public lands)
Department of Transportation—
Coast Guard (bridges, navigation)
Environmental Protection Agency (pollution effects)
National Aeronautics and Space Administration (remote sensing)

REDEVELOPMENT AND CONSTRUCTION IN BUILT-UP AREAS

Department of Commerce—
Economic Development Administration (designated areas)
Department of Housing and Urban Development
Department of the Interior—
Office of Land Use and Water Planning
Department of Transportation
Environmental Protection Agency
General Services Administration
Office of Economic Opportunity

DENSITY AND CONGESTION MITIGATION

Department of Health, Education, and Welfare
Department of Housing and Urban Development
Department of the Interior—
Office of Land Use and Water Planning
Bureau of Outdoor Recreation
Department of Transportation
Environmental Protection Agency

NEIGHBORHOOD CHARACTER AND CONTINUITY

Department of Health, Education, and Welfare
Department of Housing and Urban Development
National Endowment for the Arts
Office of Economic Opportunity

IMPACTS ON LOW-INCOME POPULATIONS

Department of Commerce—
Economic Development Administration (designated areas)
Department of Health, Education, and Welfare

Department of Housing and Urban Development
Office of Economic Opportunity

HISTORIC, ARCHITECTURAL, AND ARCHEOLOGICAL PRESERVATION

Advisory Council on Historic Preservation
Department of Housing and Urban Development
Department of the Interior—
  National Park Service
  Bureau of Land Management (public lands)
  Bureau of Indian Affairs (Indian lands)
General Services Administration
National Endowment for the Arts

SOIL AND PLANT CONSERVATION AND HYDROLOGY

Department of Agriculture—
  Soil Conservation Service
  Agricultural Service
  Forest Service
Department of Commerce—
  National Oceanic and Atmospheric Administration
Department of Defense—
  Army Corps of Engineers (dredging, aquatic plants)
Department of Health, Education, and Welfare
Department of the Interior—
  Bureau of Land Management
  Bureau of Sport Fisheries and Wildlife
  Geological Survey
  Bureau of Reclamation
Environmental Protection Agency
National Aeronautics and Space Administration (remote sensing)
River Basin Commissions (as geographically appropriate)
Water Resources Council

OUTDOOR RECREATION

Department of Agriculture—
  Forest Service
  Soil Conservation Service
Department of Defense—
  Army Corps of Engineers
Department of Housing and Urban Development (urban areas)
Department of the Interior—
  Bureau of Land Management
  National Park Service
  Bureau of Outdoor Recreation
  Bureau of Sport Fisheries and Wildlife

Bureau of Indian Affairs
Environmental Protection Agency
National Aeronautics and Space Administration (remote sening)
River Basin Commissions (as geographically appropriate)
Water Resources Council

APPENDIX III—OFFICES WITHIN FEDERAL AGENCIES AND FEDERAL-STATE AGENCIES FOR INFORMATION REGARDING THE AGENCIES' NEPA ACTIVITIES AND FOR RECEIVING OTHER AGENCIES' IMPACT STATEMENTS FOR WHICH COMMENTS ARE REQUESTED

ADVISORY COUNCIL ON HISTORIC PRESERVATION

Office of Architectural and Environmental Preservation, Advisory Council on Historic Preservation, Suite 430, 1522 K Street, N.W., Washington, D.C. 20005 254–3974

DEPARTMENT OF AGRICULTURE [1]

Office of the Secretary, Attn: Coordinator Environmental Quality Activities, U.S. Department of Agriculture, Washington, D.C. 20250 447–3965

APPALACHIAN REGIONAL COMMISSION

Office of the Alternate Federal Co-Chairman, Appalachian Regional Commission, 1666 Connecticut Avenue, N.W., Washington, D.C. 20235 967–4103

DEPARTMENT OF THE ARMY (CORPS OF ENGINEERS

Executive Director of Civil Works, Office of the Chief of Engineers, U.S. Army Corps of Engineers, Washington, D.C. 20314 693–7168

ATOMIC ENERGY COMMISSION

For nonregulatory matters: Office of Assistant General Manager for Biomedical and Environmental Research and Safety Programs, Atomic Energy Com-

[1] Requests for comments or information from individual units of the Department of Agriculture, e.g., Soil Conservation Service, Forest Service, etc. should be sent to the Office of the Secretary, Department of Agriculture, at the address given above.

mission, Washington, D.C. 20545 973–3208

For regulatory matters: Office of the Assistant Director for Environmental Projects, Atomic Energy Commission, Washington, D.C. 20545 973–7531

DEPARTMENT OF COMMERCE

Office of the Deputy Assistant Secretary for Environmental Affairs, U.S. Department of Commerce, Washington, D.C. 20230 967–4335

DEPARTMENT OF DEFENSE

Office of the Assistant Secretary for Defense (Health and Environment), U.S. Department of Defense, Room 3E172, The Pentagon, Washington, D.C. 20301 697–2111

DELAWARE RIVER BASIN COMMISSION

Office of the Secretary, Delaware River Basin Commission, Post Office Box 360, Trenton, N.J. 08603 (609) 883–9500

ENVIRONMENTAL PROTECTION AGENCY [2]

Director, Office of Federal Activities, Environmental Protection Agency, 401 M Street, S.W., Washington, D.C. 20460 755–0777

For all other EPA consultations, contact the Regional Administrator in whose area the proposed action (e.g., highway or water resource construction projects) will take place. The Regional Administrators will coordinate the EPA review. Addresses of the Regional Administrators, and the areas covered by their regions are as follows:

Regional Administrator, I,
U.S. Environmental Protection Agency
Room 2303, John F. Kennedy
Federal Bldg., Boston, Mass. 02203,
(617) 223–7210

Regional Administrator, II,
U.S. Environmental Protection Agency
Room 908, 26 Federal Plaza
New York, New York 10007
(212) 264–2525

Regional Administrator, III,
U.S. Environmental Protection Agency
Curtis Bldg., 6th & Walnut Sts.
Philadelphia, Pa. 19106
(215) 597–9801

Regional Administrator, IV,
U.S. Environmental Protection Agency
1421 Peachtree Street
N.E., Atlanta, Ga. 30309
(404) 526–5727

Regional Administrator, V,
U.S. Environmental Protection Agency
1 N. Wacker Drive
Chicago, Illinois 60606
(312) 353–5250

Regional Administrator, VI,
U.S. Environmental Protection Agency
1600 Patterson Street
Suite 1100
Dallas, Texas 75201
(214) 749–1962

Regional Administrator, VII,
U.S. Environmental Protection Agency
1735 Baltimore Avenue
Kansas City, Missouri 64108
(816) 374–5493

Regional Administrator, VIII,
U.S. Environmental Protection Agency
Suite 900, Lincoln Tower
1860 Lincoln Street
Denver, Colorado 80203
(303) 837–3895

Regional Administrator, IX,
U.S. Environmental Protection Agency
100 California Street
San Francisco, California 94111
(415) 556–2320

Regional Administrator, X,
U.S. Environmental Protection Agency
1200 Sixth Avenue
Seattle, Washington 98101
(206) 442–1220

Connecticut, Maine, Massachusetts, New Hampshire, Rhode Island, Vermont

New Jersey, New York, Puerto Rico, Virgin Islands

Delaware, Maryland, Pennsylvania, Virginia, West Virginia, District of Columbia

[2] Contact the Office of Federal Activities for environmental statements concerning legislation, regulations, national program proposals or other major policy issues.

Council on Environmental Quality Guidelines 139

Alabama, Florida, Georgia, Kentucky, Mississippi, North Carolina, South Carolina, Tennessee

Illinois, Indiana, Michigan, Minnesota, Ohio, Wisconsin

Arkansas, Louisiana, New Mexico, Texas, Oklahoma

Iowa, Kansas, Missouri, Nebraska

Colorado, Montana, North Dakota, South Dakota, Utah, Wyoming

Arizona, California, Hawaii, Nevada, American Samoa, Guam, Trust Territories of Pacific Islands, Wake Island

Alaska, Idaho, Oregon, Washington

Director, Office of Federal Activities, Environmental Protection Agency, 401 M Street, S.W., Washington, D.C. 20460 755-0777

FEDERAL POWER COMMISSION

Commission's Advisor on Environmental Quality, Federal Power Commission, 825 N. Capitol Street, N.E., Washington, D.C. 20426 386-6084

GENERAL SERVICES ADMINISTRATION

Office of Environmental Affairs, Office of the Deputy Administrator for Special Projects, General Services Administration, Washington, D.C. 20405 343-4161

GREAT LAKES BASIN COMMISSION

Office of the Chairman, Great Lakes Basin Commission, 3475 Plymouth Road, P.O. Box 999, Ann Arbor, Michigan 48105 (313) 769-7431

DEPARTMENT OF HEALTH, EDUCATION AND WELFARE [3]

Region I:
Regional Environmental Officer
U.S. Department of Health, Education and Welfare
Room 2007B
John F. Kennedy Center
Boston, Massachusetts 02203 (617) 223-6837
Region II:
Regional Environmental Officer
U.S. Department of Health, Education and Welfare
Federal Building
26 Federal Plaza
New York, New York 10007 (212) 264-1308
Region III:
Regional Environmental Officer
U.S. Department of Health, Education and Welfare
P.O. Box 13716
Philadelphia, Pennsylvania 19101 (215) 597-6498
Region IV:
Regional Environmental Officer
U.S. Department of Health, Education and Welfare
Room 404
50 Seventh Street, N.E.
Atlanta, Georgia 30323 (404) 526-5817
Region V:
Regional Environmental Officer
U.S. Department of Health, Education and Welfare

[3] Contact the Office of Environmental Affairs for information on HEW's environmental statements concerning legislation, regulations, national program proposals or other major policy issues, and for all requests for HEW comment on impact statements of other agencies.
For information with respect to HEW actions occurring within the jurisdiction of the Departments' Regional Directors, contact the appropriate Regional Environmental Officer:

Room 712, New Post Office Building
433 West Van Buren Street
Chicago, Illinois 60607 (312) 353-1644

Region VI:
Regional Environmental Officer
U.S. Department of Health, Education and Welfare
1114 Commerce Street
Dallas, Texas 75202 (214) 749-2236

Region VII:
Regional Environmental Officer
U.S. Department of Health, Education and Welfare
601 East 12th Street
Kansas City, Missouri 64106 (816) 374-3584

Region VIII:
Regional Environmental Officer
U.S. Department of Health, Education and Welfare
9017 Federal Building
19th and Stout Streets
Denver, Colorado 80202 (303) 837-4178

Region IX:
Regional Environmental Officer
U.S. Department of Health, Education and Welfare
50 Fulton Street
San Francisco, California 94102 (415) 556-1970

Region X:
Regional Environmental Officer
U.S. Department of Health, Education and Welfare
Arcade Plaza Building
1321 Second Street
Seattle, Washington 98101 (206) 442-0490

Office of Environmental Affairs, Office of the Assistant Secretary for Administration and Management, Department of Health, Education and Welfare, Washington, D.C. 20202 963-4456

DEPARTMENT OF HOUSING AND URBAN DEVELOPMENT [4]

Regional Administrator I,
Environmental Clearance Officer

[4] Contact the Director with regard to environmental impacts of legislation, policy statements, program regulations and procedures, and precedent-making project decisions. For all other HUD consultation, contact the HUD Regional Administrator in whose jurisdiction the project lies, as follows:

U.S. Department of Housing and Urban Development
Room 405, John F. Kennedy Federal Building
Boston, Mass. 02203 (617) 223-4066

Regional Administrator II,
Environmental Clearance Officer
U.S. Department of Housing and Urban Development
26 Federal Plaza
New York, New York 10007 (212) 264-8068

Regional Administrator III,
Environmental Clearance Officer
U.S. Department of Housing and Urban Development
Curtis Building, Sixth and Walnut Streets
Philadelphia, Pennsylvania 19106 (215) 597-2560

Regional Administrator IV,
Environmental Clearance Officer
U.S. Department of Housing and Urban Development
Peachtree-Seventh Building
Atlanta, Georgia 30323 (404) 526-5585

Regional Administrator V,
Environmental Clearance Officer
U.S. Department of Housing and Urban Development
360 North Michigan Avenue
Chicago, Illinois 60601 (312) 353-5680

Regional Administrator VI,
Environmental Clearance Officer
U.S. Department of Housing and Urban Development
Federal Office Building, 819 Taylor Street
Fort Worth, Texas 76102 (817) 334-2867

Regional Administrator VII,
Environmental Clearance Officer
U.S. Department of Housing and Urban Development
911 Walnut Street
Kansas City, Missouri 64106 (816) 374-2661

Regional Administrator VIII,
Environmental Clearance Officer
U.S. Department of Housing and Urban Development
Samsonite Building, 1051 South Broadway
Denver, Colorado 80209 (303) 837-4061

Regional Administrator IX,
  Environmental Clearance Officer
  U.S. Department of Housing and Urban Development
  450 Golden Gate Avenue, Post Office Box 36003
  San Francisco, California 94102
  (415) 556-4752
Regional Administrator X,
  Environmental Clearance Officer
  U.S. Department of Housing and Urban Development
  Room 226, Arcade Plaza Building
  Seattle, Washington 98101 (206) 583-5415
Director, Office of Community and Environmental Standards, Department of Housing and Urban Development, Room 7206, Washington, D.C. 20410
755-5980

DEPARTMENT OF THE INTERIOR [5]

Director, Office of Environmental Project Review, Department of the Interior, Interior Building, Washington, D.C. 20240 343-3891

INTERSTATE COMMERCE COMMISSION

Office of Proceedings, Interstate Commerce Commission, Washington, D.C. 20423
343-6167

DEPARTMENT OF LABOR

Assistant Secretary for Occupational Safety and Health, Department of Labor, Washington, D.C.
961-3405

MISSOURI RIVER BASINS COMMISSION

Office of the Chairman, Missouri River Basins Commission, 10050 Regency Circle, Omaha, Nebraska 68114
(402) 397-5714

NATIONAL AERONAUTICS AND SPACE ADMINISTRATION

Office of the Comptroller, National Aeronautics and Space Administration, Washington, D.C. 20546
755-8440

[5] Requests for comments or information from individual units of the Department of the Interior should be sent to the Office of Environmental Project Review at the address given above.

NATIONAL CAPITAL PLANNING COMMISSION

Office of Environmental Affairs, Office of the Executive Director, National Capital Planning Commission, Washington, D.C. 20576
382-7200

NATIONAL ENDOWMENT FOR THE ARTS

Office of Architecture and Environmental Arts Program, National Endowment for the Arts, Washington, D.C. 20506
382-5765

NEW ENGLAND RIVER BASINS COMMISSION

Office of the Chairman, New England River Basins Commission, 55 Court Street, Boston, Mass. 02108
(617) 223-6244

OFFICE OF ECONOMIC OPPORTUNITY

Office of the Director, Office of Economic Opportunity, 1200 19th Street, N.W., Washington, D. C. 20506
254-6000

OHIO RIVER BASIN COMMISSION

Office of the Chairman, Ohio River Basin Commission, 36 East 4th Street, Suite 208-20, Cincinnati, Ohio 45202
(513) 684-3831

PACIFIC NORTHWEST RIVER BASINS COMMISSION

Office of the Chairman, Pacific Northwest River Basins Commission, 1 Columbia River, Vancouver, Washington 98660
(206) 695-3606

SOURIS-RED-RAINY RIVER BASINS COMMISSION

Office of the Chairman, Souris-Red-Rainy River Basins Commission, Suite 6, Professional Building, Holiday Mall, Moorhead, Minnesota 56560
(701) 237-5227

DEPARTMENT OF STATE

Office of the Special Assistant to the Secretary for Environmental Affairs, Department of State, Washington, D.C. 20520
632-7964

SUSQUEHANNA RIVER BASIN COMMISSION

Office of the Executive Director, Susquehanna River Basin Commission, 5012 Lenker Street, Mechanicsburg, Pa. 17055 (717) 737-0501

TENNESSEE VALLEY AUTHORITY

Office of the Director of Environmental Research and Development, Tennessee Valley Authority, 720 Edney Building, Chattanooga, Tennessee 37401 (615) 755-2002

DEPARTMENT OF TRANSPORTATION [6]

U.S. Coast Guard

Office of Marine Environment and Systems, U.S. Coast Guard, 400 7th Street, S.W., Washington, D.C. 20590, 426-2007

Federal Aviation Administration

Office of Environmental Quality, Federal Aviation Administration, 800 Independence Avenue, S.W., Washington, D.C. 20591, 426-8406

Federal Highway Administration

Office of Environmental Policy, Federal Highway Administration, 400 7th Street, S.W., Washington, D.C. 20590, 426-0351

Federal Railroad Administration

Office of Policy and Plans, Federal Railroad Administration, 400 7th Street, S.W., Washington, D.C. 20590, 426-1567

Urban Mass Transportation Administration

Office of Program Operations, Urban Mass Transportation Administration,

[6] Contact the Office of Environmental Quality, Department of Transportation, for information on DOT's environmental statements concerning legislation, regulations, national program proposals, or other major policy issues.

For information regarding the Department of Tranportation's other environmental statements, contact the national office for the appropriate administration:

400 7th Street, S.W., Washington, D.C. 20590, 426-4020

For other administrations not listed above, contact the Office of Environment Quality, Department of Transportation, at the address given above.

For comments on other agencies' environmental statements, contact the appropriate administration's regional office. If more than one administration within the Department of Transportation is to be requested to comment, contact the Secretarial Representative in the appropriate Regional Office for coordination of the Department's comments:

SECRETARIAL REPRESENTATIVE

Region I Secretarial Representative, U.S. Department of Transportation, Transportation Systems Center, 55 Broadway, Cambridge, Massachusetts 02142 (617) 494-2709

Region II Secretarial Representative, U.S. Department of Transportation, 26 Federal Plaza, Room 1811, New York, New York 10007 (212) 264-2672

Region III Secretarial Representative, U.S. Department of Transportation, Mall Building, Suite 1214, 325 Chestnut Street, Philadelphia, Pennsylvania 19106 (215) 597-0407

Region IV Secretarial Representative, U.S. Department of Transportation, Suite 515, 1720 Peachtree Rd., N.W. Atlanta, Georgia 30309 (404) 526-3738

Region V Secretarial Representative, U.S. Department of Transportation, 17th Floor, 300 S. Wacker Drive, Chicago, Illinois 60606 (312) 353-4000

Region VI Secretarial Representative, U.S. Department of Transportation, 9-C-18 Federal Center, 1100 Commerce Street, Dallas, Texas 75202 (214) 749-1851

Region VII Secretarial Representative, U.S. Department of Transportation, 601 E. 12th Street, Room 634, Kansas City, Missouri 64106 (816) 374-2761

Region VIII Secretarial Representative, U.S. Department of Transportation, Prudential Plaza, Suite 1822, 1050 17th Street, Denver, Colorado 80225 (303) 837-3242

Region IX Secretarial Representative, U.S. Department of Transportation, 450 Golden Gate Avenue, Box 36133,

San Francisco, California 94102 (415) 556-5961

Region X Secretarial Representative, U.S. Department of Transportation, 1321 Second Avenue, Room 507, Seattle, Washington 98101 (206) 442-0590

FEDERAL AVIATION ADMINISTRATION

New England Region, Office of the Regional Director, Federal Aviation Administration, 154 Middlesex Street, Burlington, Massachusetts 01803 (617) 272-2350

Eastern Region, Office of the Regional Director, Federal Aviation Administration, Federal Building, JFK International Airport, Jamaica, New York 11430 (212) 995-3333

Southern Region, Office of the Regional Director, Federal Aviation Administration, P.O. Box 20636, Atlanta, Georgia 30320 (404) 526-7222

Great Lakes Region, Office of the Regional Director, Federal Aviation Administration, 2300 East Devon, Des Plaines, Illinois 60018 (312) 694-4500

Southwest Region, Office of the Regional Director, Federal Aviation Administration, P.O. Box 1689, Fort Worth, Texas 76101 (817) 624-4911

Central Region, Office of the Regional Director, Federal Aviation Administration, 601 E. 12th Street, Kansas City, Missouri 64106 (816) 374-5626

Rocky Mountain Region, Office of the Regional Director, Federal Aviation Administration, Park Hill Station, P.O. Box 7213, Denver, Colorado 80207 (303) 837-3646

Western Region, Office of the Regional Director, Federal Aviation Administration, P.O. Box 92007, WorldWay Postal Center, Los Angeles, California 90009 (213) 536-6427

Northwest Region, Office of the Regional Director, Federal Aviation Administration, FAA Building, Boeing Field, Seattle, Washington 98108 (206) 767-2780

FEDERAL HIGHWAY ADMINISTRATION

Region 1, Regional Administrator, Federal Highway Administration, 4 Normanskill Boulevard, Delmar, New York 12054 (518) 472-6476

Region 3, Regional Administrator, Federal Highway Administration, Room 1621, George H. Fallon Federal Office Building, 31 Hopkins Plaza, Baltimore, Maryland 21201 (301) 962-2361

Region 4, Regional Administration, Federal Highway Administration, Suite 200, 1720 Peachtree Road, N.W., Atlanta, Georgia 30309 (404) 526-5078

Region 5, Regional Administrator, Federal Highway Administration, Dixie Highway, Homewood, Illinois 60430 (312) 799-6300

Region 6, Regional Administrator, Federal Highway Administration, 819 Taylor Street, Fort Worth, Texas 76102 (817) 334-3232

Region 7, Regional Administrator, Federal Highway Administration, P.O. Box 7186, Country Club Station, Kansas City, Missouri 64113 (816) 361-7563

Region 8, Regional Administrator, Federal Highway Administration, Room 242, Building 40, Denver Federal Center, Denver, Colorado 80225

Region 9, Regional Administrator, Federal Highway Administration, 450 Golden Gate Avenue, Box 36096, San Francisco, California 94102 (415) 556-3895

Region 10, Regional Administrator, Federal Highway Administration, Room 412, Mohawk Building, 222 S.W. Morrison Street, Portland, Oregon 97204 (503) 221-2065

URBAN MASS TRANSPORTATION ADMINISTRATION

Region I, Office of the UMTA Representative, Urban Mass Transportation Administration, Transportation Systems Center, Technology Building, Room 277, 55 Broadway, Boston, Massachusetts 02142 (617) 494-2055

Region II, Office of the UMTA Representative, Urban Mass Transportation Administration, 26 Federal Plaza, Suite 1809, New York, New York 10007 (212) 264-8162

Region III, Office of the UMTA Representative, Urban Mass Transportation Administration, Mall Building, Suite 1214, 325 Chestnut Street, Philadelphia, Pennsylvania 19106 (215) 597-0407

Region IV, Office of UMTA Representative, Urban Mass Transportation Administration, 1720 Peachtree Road, Northwest, Suite 501, Atlanta, Georgia 30309 (404) 526-3948

Region V, Office of the UMTA Representative, Urban Mass Transportation Administration, 300 South Wacker Drive, Suite 700, Chicago, Illinois 60606 (312) 353-6005

Region VI, Office of the UMTA Representative, Urban Mass Transportation Administration, Federal Center, Suite 9E24, 1100 Commerce Street, Dallas, Texas 75202 (214) 749-7322

Region VII, Office of the UMTA Representative, Urban Mass Transportation Administration, c/o FAA Management Systems Division, Room 1564D, 601 East 12th Street, Kansas City, Missouri 64106 (816) 374-5567

Region VIII, Office of the UMTA Representative, Urban Mass Transportation Administration, Prudential Plaza, Suite 1822, 1050 17th Street, Denver, Colorado 80202 (303) 837-3242

Region IX, Office of the UMTA Representative, Urban Mass Transportation Administration, 450 Golden Gate Avenue, Box 36125, San Francisco, California 94102 (415) 556-2884

Region X, Office of the UMTA Representative, Urban Mass Transportation Administration, 1321 Second Avenue, Suite 5079, Seattle, Washington (206) 442-0590

DEPARTMENT OF THE TREASURY

Office of Assistant Secretary for Administration, Department of the Treasury, Washington, D.C. 20220 964-5391

UPPER MISSISSIPPI RIVER BASIN COMMISSION

Office of the Chairman, Upper Mississippi River Basin Commission, Federal Office Building, Fort Snelling, Twin Cities, Minnesota 55111 (612) 725-4690

WATER RESOURCES COUNCIL

Office of the Associate Director, Water Resources Council, 2120 L Street, N.W., Suite 800, Washington, D.C. 20037 254-6442

APPENDIX IV—STATE AND LOCAL AGENCY REVIEW OF IMPACT STATEMENTS

1. OMB Circular No. A-95 through its system of clearinghouses provides a means for securing the views of State and local environmental agencies, which can assist in the preparation of impact statements. Under A-95, review of the proposed project in the case of federally assisted projects (Part I of A-95) generally takes place prior to the preparation of the impact statement. Therefore, comments on the environmental effects of the proposed project that are secured during this stage of the A-95 process represent inputs to the environmental impact statement.

2. In the case of direct Federal development (Part II of A-95), Federal agencies are required to consult with clearinghouses at the earliest practicable time in the planning of the project or activity. Where such consultation occurs prior to completion of the draft impact statement, comments relating to the environmental effects of the proposed action would also represent inputs to the environmental impact statement.

3. In either case, whatever comments are made on environmental effects of proposed Federal or federally assisted projects by clearinghouses, or by State and local environmental agencies through clearinghouses, in the course of the A-95 review should be attached to the draft impact statement when it is circulated for review. Copies of the statement should be sent to the agencies making such comments. Whether those agencies then elect to comment again on the basis of the draft impact statement is a matter to be left to the discretion of the commenting agency depending on its resources, the significance of the project, and the extent to which its earlier comments were considered in preparing the draft statement.

4. The clearinghouses may also be used, by mutual agreement, for securing reviews of the draft environmental impact statement. However, the Federal agency may wish to deal directly with appropriate State or local agencies in the review of impact statements because the clearinghouses may be unwilling or unable to handle this phase of the process. In some cases, the Governor may have designated a specific agency, other than the clearinghouse, for securing reviews of impact statements. In any case, the clearinghouses should be sent copies of the impact statement.

5. To aid clearinghouses in coordinating State and local comments, draft statements should include copies of State and local agency comments made earlier under the A-95 process and should indicate on the summary sheet those other agencies from which comments have been requested, as specified in Appendix I of the CEQ Guidelines.

[FR Doc. 73-15783 Filed 7-31-73; 8:45 am]

ERRATA SHEET

The *Federal Register* publication of the Council on Environmental Quality Guidelines contains the following typesetting errors.

Table of Contents:	(1) Section 1500.9 should read "Review of draft environmental statements by Federal, Federal-State, State, and local agencies, and by the public."
	(2) The word "Sec." after Section 1500.14 should be deleted.
Section 1500.6(a):	"enviornmental" should read "environmental".
Section 1500.6(c):	"(see § 1500.5 ii)" should read "(see § 1500.5(a)(2))".
Section 1500.6(c)(iii):	"paragraph (C)(4)(ii)" should read "paragraph (ii)".
Section 1500.6(e):	"§ 1500.6(c)(4)(ii)" should read "§ 1500.6(c)(ii)".
Section 1500.7(a):	"§ 1500.6(c)(c)(ii)" should read "§ 1500.6(c)(ii)".
Section 1500.8(a)(1):	"see paragraph (a)(1)(3)(ii)" should read "see paragraph (3)(ii)".
Section 1500.9(a):	"(a) *Federal agency review.* (1) *in*" should read "(a) *Federal agency review: in*".
Section 1500.9(c):	"securning" should read "securing".
Effective date:	The effective date should read January 28, 1974, not January 28, 1973.
Appendix I:	"WATER RESOURCES COUNCIL" appearing after the heading *Weather Modification* should be aligned with the left hand margin and read "Water Resources Council".
	"PROTECTION OF ENVIRONMENTALLY CRITICAL AREAS—FLOODPLAINS, WETLANDS, BEACHES AND DUNES, UNSTABLE SOILS, STEEP SLOPES, AQUIFER RECHARGE AREAS, ETC." should read *"Protection of Environmentally Critical Areas—Floodplains, Wetlands, Beaches and Dunes, Unstable Soils, Steep Slopes, Aquifer Recharge Areas, Etc."*
	"LAND USE IN COASTAL AREAS" should read *"Land Use in Coastal Areas"*
	"REDEVELOPMENT AND CONSTRUCTION IN BUILT-UP AREAS" should read *"Redevelopment and Construction In Built-Up Areas"*
	"DENSITY AND CONGESTION MITIGATION" should read *"Density and Congestion Mitigation"*
	"NEIGHBORHOOD CHARACTER AND CONTINUITY" should read *"Neighborhood Character and Continuity"*
	"IMPACTS ON LOW INCOME POPULATIONS" should read *"Impacts on Low-Income Populations"*
	"HISTORIC, ARCHITECTURAL, AND ARCHEOLOGICAL PRESERVATION" should read *"Historic, Architectural, and Archeological Preservation"*
	"SOIL AND PLANT CONSERVATION AND HYDROLOGY" should read *"Soil And Plant Conservation and Hydrology"*

APPENDIX

3

Office of Management
Circular A-95,
Revised

EXECUTIVE OFFICE OF THE PRESIDENT
OFFICE OF MANAGEMENT AND BUDGET
WASHINGTON, D.C. 20503

February 9, 1971

CIRCULAR NO. A-95
Revised

TO THE HEADS OF EXECUTIVE DEPARTMENTS
AND ESTABLISHMENTS

SUBJECT: Evaluation, review, and coordination of Federal and federally assisted programs and projects

1. *Purpose.* This Circular furnishes guidance to Federal agencies for added cooperation with State and local governments in the evaluation, review, and coordination of Federal assistance programs and projects. The Circular promulgates regulations (Attachment A) which provide, in part, for:

    a. Encouraging the establishment of a project notification and review system to facilitate coordinated planning on an intergovernmental basis for certain Federal assistance programs in furtherance of section 204 of the Demonstration Cities and Metropolitan Development Act of 1966 and Title IV of the Intergovernmental Cooperation Act of 1968 (Attachment B).

    b. Coordination of direct Federal development programs and projects with State, regional, and local planning and programs pursuant to Title IV of the Intergovernmental Cooperation Act of 1968.

    c. Securing the comments and views of State and local agencies which are authorized to develop and enforce environmental standards on certain Federal or federally assisted projects affecting the environment pursuant to section 102(2) (C) of the National Environmental Policy Act of 1969 (Attachment C) and regulations of the Council on Environmental Quality.

    This Circular supersedes Circular No. A-95, dated July 24, 1969, as amended by Transmittal Memorandum No. 1, dated December 27, 1969. It will become effective April 1, 1971.

2. *Basis.* This Circular has been prepared pursuant to:

    a. Section 401(a) of the Intergovernmental Cooperation Act of 1968 which provides, in part, that

"The President shall . . . establish rules and regulations governing the formulation, evaluation, and review of Federal programs and projects having a significant impact on area and community development . . ."

and the President's Memorandum of November 8, 1968, to the Director of the Bureau of the Budget ("Federal Register," Vol. 33, No. 221, November 13, 1968) which provides:

"By virtue of the authority vested in me by section 301 of title 3 of the United States Code and section 401(a) of the Intergovernmental Cooperation Act of 1968 (Public Law 90-577), I hereby delegate to you the authority vested in the President to establish the rules and regulations provided for in that section governing the formulation, evaluation, and review of Federal programs and projects having a significant impact on area and community development, including programs providing Federal assistance to the States and localities, to the end that they shall most effectively serve these basic objectives.

"In addition, I expect the Bureau of the Budget to generally coordinate the actions of the departments and agencies in exercising the new authorizations provided by the Intergovernmental Cooperation Act, with the objective of consistent and uniform action by the Federal Government."

b. Title IV, section 403, of the Intergovernmental Cooperation Act of 1968 which provides that:

"The Bureau of the Budget, or such other agency as may be designated by the President, shall prescribe such rules and regulations as are deemed appropriate for the effective administration of this Title."

c. Section 204 (c) of the Demonstration Cities and Metropolitan Development Act of 1966 which provides that:

"The Bureau of the Budget, or such other agency as may be designated by the President, shall prescribe such rules and regulations as are deemed appropriate for the effective administration of this section," and

d. Reorganization Plan No. 2 of 1970 and Executive Order No. 11541 of July 1, 1970, which vest all functions of the Bureau of the Budget or the Director of the Bureau of the Budget in the Director of the Office of Management and Budget.

3. *Coverage.* The regulations promulgated by this Circular (Attachment A) will have applicability to:

a. Under Part I, all projects (or significant changes thereto) for which Federal assistance is being sought under the programs listed in Attachment D. Limitations and provision for exceptions are noted therein.

b. Under Part II, all direct Federal development activities, including the acquisition, use, and disposal of Federal real property.

c. Under Part III, all Federal programs requiring, by statute or administrative regulation, a State plan as a condition of assistance.

d. Under Part IV, all Federal programs providing assistance to State, local, and regional projects and activities that are planned on a multijurisdictional basis.

4. *Inquiries.* Inquiries concerning this Circular may be addressed to the Office of Management and Budget, Washington, D. C. 20503, telephone (202) 305-3031 (Government dial code 103-3031).

GEORGE P. SHULTZ
Director

Attachments

ATTACHMENT A
Circular No. A-95
Revised

REGULATIONS UNDER SECTION 204 OF THE DEMONSTRATION CITIES AND METROPOLITAN DEVELOPMENT ACT OF 1966, TITLE IV OF THE INTERGOVERNMENTAL COOPERATION ACT OF 1968, AND SECTION 102 (2) (C) OF THE NATIONAL ENVIRONMENTAL POLICY ACT OF 1969

PART I: PROJECT NOTIFICATION AND REVIEW SYSTEM

1. *Purpose.* The purpose of this Part is to:

a. Further the policies and directives of Title IV of the Intergovernmental Cooperation Act of 1968 by encouraging the establishment of a network of State, regional, and metropolitan planning and development clearinghouses which will aid in the coordination of Federal or federally assisted projects and programs with State, regional, and local planning for orderly growth and development;

b. Implement the requirements of section 204 of the Demonstration Cities and Metropolitan Development Act of 1966 for metropolitan areas within that network;

Office of Management Circular A-95, Revised 151

c. Implement, in part, requirements of section 102(2) (C) of the National Environmental Policy Act of 1969, which require State and local views of the environmental impact of Federal or federally assisted projects.

d. Encourage, by means of early contact between applicants for Federal assistance and State and local governments and agencies, an expeditious process of intergovernmental coordination and review of proposed projects.

2. *Notification.*

a. Any agency of State or local government or any organization or individual undertaking to apply for assistance to a project under a Federal program listed in Attachment D will be required to notify the planning and development clearinghouse of the State (or States) and the region, if there is one, or of the metropolitan area in which the project is to be located, of its intent to apply for assistance. Notification will be accompanied by a summary description of the project for which assistance will be sought. The summary description will contain the following information:

(1) Identity of the applicant agency, organization, or individual.

(2) The geographic location of the project to be assisted.

(3) A brief description of the proposed project by type, purpose, general size or scale, estimated cost, beneficiaries, or other characteristics which will enable the clearinghouse to identify agencies of State or local government having plans, programs, or projects that might be affected by the proposed projects.

(4) A brief statement of whether or not an environmental impact statement is required and, if so, an indication of the nature and extent of environmental impact anticipated.

(5) The Federal program and agency under which assistance will be sought as indicated in the *Catalog of Federal Domestic Assistance* (April 1970 and subsequent editions).

(6) The estimated date by which time the applicant expects to formally file an application.

Many clearinghouses have developed notification forms and instructions. Applicants are urged to contact their clearinghouses for such information in order to expedite clearinghouse review.

b. In order to assure maximum time for effective coordination and so as not to delay the timely submission of the completed application to the Federal agency, such notifications should be sent at the earliest feasible time.

3. *Clearinghouse functions.* Clearinghouse functions include:

a. Evaluating the significance of proposed Federal or federally assisted projects to State, areawide or local plans and programs, as appropriate.

b. Receiving and disseminating project notifications to appropriate State agencies in the case of the State clearinghouse and to appropriate local governments and agencies in the case of regional or metropolitan clearinghouses; and providing liaison, as may be necessary, between such agencies or bodies and the applicant.

c. Assuring, pursuant to section 102(2) (C) of the National Environmental Policy Act of 1969, that appropriate State, metropolitan, regional, or local agencies which are authorized to develop and enforce environmental standards are informed of and are given opportunity to review and comment on the environmental significance of proposed projects for which Federal assistance is sought.

d. Providing, pursuant to Part II of these regulations, liaison between Federal agencies contemplating direct Federal development projects and the State or areawide agencies or local governments having plans or programs that might be affected by the proposed project.

4. *Consultation and review.*

a. State, metropolitan, and regional clearinghouses may have a period of 30 days after receipt of a project notification in which to inform State agencies, other local or regional bodies, etc., that may be affected by the project (including agencies authorized to develop and enforce environmental standards) and to arrange, as may be necessary, to consult with the applicant on the proposed project.

b. During this period and during the period in which the application is being completed, the clearinghouse may work with the applicant in the resolution of any problems raised by the proposed project.

c. Clearinghouses may have, if necessary, an additional 30 days to review the completed application and to transmit to the applicant any comments or recommendations the clearinghouse (or others) may have.

d. In the case of a project for which Federal assistance is sought by a special purpose unit of government, clearinghouses will assure that any unit of general local government, having jurisdiction over the area in which the project is to be located, has opportunity to confer, consult, and comment upon the project and the application.

e. Applicants will include with the completed application as submitted to the Federal agency:

(1) Any comments and recommendations made by or through clearinghouses, along with a statement that such comments have been considered prior to submission of the application; *or*

(2) A statement that the procedures outlined in this section have been followed and that no comments or recommendations have been received.

f. Where regional or metropolitan areas are contiguous, coordinative arrangements should be established between the clearinghouses in such areas to assure that projects in one area which may have an impact on the development of a contiguous area are jointly studied. Any comments and recommendations made by or through a clearinghouse in one area on a project in a contiguous area will accompany the application for assistance to that project.

5. *Subject matter of comments and recommendations.* Comments and recommendations made by or through clearinghouses with respect to any project are for the purpose of assuring maximum consistency of such project with State, regional and local comprehensive plans. They are also intended to assist the Federal agency (or State agency, in the case of projects for which the State under certain Federal grants has final project approval) administering such a program in determining whether the project is in accord with applicable Federal law. Comments or recommendations, as may be appropriate, may include information about:

   a. The extent to which the project is consistent with or contributes to the fulfillment of comprehensive planning for the State, region, metropolitan area, or locality.

   b. The extent to which the project contributes to the achievement of State, regional, metropolitan, and local objectives as specified in section 401(a) of the Intergovernmental Cooperation Act of 1968, as follows:

   (1) Appropriate land uses for housing, commercial, industrial, governmental, institutional, and other purposes;

   (2) Wise development and conservation of natural resources, including land, water, minerals, wildlife, and others;

   (3) Balanced transportation systems, including highway, air, water, pedestrian, mass transit, and other modes for the movement of people and goods;

   (4) Adequate outdoor recreation and open space;

   (5) Protection of areas of unique natural beauty, historical and scientific interest;

   (6) Properly planned community facilities, including utilities for the supply of power, water, and communications, for the safe disposal of wastes, and for other purposes; and

   (7) Concern for high standards of design.

   c. As provided under section 102(2)(C) of the National Environmental Policy Act of 1969, the extent to which the project significantly affects the environment including consideration of:

   (1) The environmental impact of the proposed project;

   (2) Any adverse environmental effects which cannot be avoided should the proposed project be implemented;

   (3) Alternatives to the proposed project;

(4) The relationship between local short term uses of man's environment and the maintenance and enhancement of long term productivity; and

(5) Any irreversible and irretrievable commitments of resources which would be involved in the proposed project or action, should it be implemented.

d. In the case of a project for which assistance is being sought by a special purpose unit of government, whether the unit of general local government having jurisdiction over the area in which the project is to be located has applied, or plans to apply for assistance for the same or similar type project. This information is necessary to enable the Federal (or State) agency to make the judgments required under section 402 of the Intergovernmental Cooperation Act of 1968.

6. *Federal agency procedures.* Federal agencies having programs covered under this Part (see Attachment D) will develop appropriate procedures for:

a. Informing potential applicants for assistance under such programs of the requirements of this Part (1) in program information materials, (2) in response to inquiries respecting application procedures, (3) in pre-application conferences, or (4) by other means which will assure earliest contact between applicant and clearinghouses.

b. Assuring that all applications for assistance under programs covered by this part have been submitted to appropriate clearinghouses for review.

c. Notifying clearinghouses within seven days of any action (approvals, disapprovals, return for amendment, etc.) taken on applications that have been reviewed by such clearinghouses. Where a State clearinghouse has assigned an identification number to an application, the Federal agency will refer to such identification number in notifying clearinghouses of actions taken on the application.

d. Assuring, in the case of an application submitted by a special purpose unit of government, where accompanying comments indicate that the unit of general local government having jurisdiction over the area in which the project is to be located has submitted or plans to submit an application for assistance for the same or a similar type project, that appropriate considerations and preferences as specified in section 402 of the Intergovernmental Cooperation Act of 1968, are accorded the unit of general local government. Where such preference cannot be so accorded, the agency shall supply, in writing, to the unit of general local government and the Office of Management and Budget its reasons therefor.

7. *HUD housing programs.* Because of the unique nature of the application and development process for the housing programs of the Depart-

ment of Housing and Urban Development, a variation of the review procedure is necessary. For HUD programs in the 14.100 series listed in Attachment D, the following procedure for review will be followed:

a. The HUD Area or Insuring Office will transmit to the appropriate State clearinghouse and metropolitan or regional clearinghouse a copy of the initial application for HUD program approval.

b. The clearinghouses will have 15 days to review the applications and to forward to the Area or Insuring Office any comments which they may have, including observations concerning the consistency of the proposed project with State and areawide development plans and identification of major environmental concerns. Processing of applications in the Area or Insuring Office will proceed concurrently with the clearinghouse review.

c. This procedure will include only applications involving new construction and will apply to:

(1) Subdivisions having 50 or more lots involving any HUD home mortgage insurance program.

(2) Multifamily projects having 100 or more dwelling units under any HUD mortgage insurance program, or under conventional or turnkey public housing programs.

(3) Mobile home courts with 100 or more spaces.

(4) College housing provided under the debt service or direct loan programs for 200 or more students. All other applications for assistance under the HUD programs in the 14.100 series listed in Attachment D are exempt from the requirements of this Circular.

8. *Reports and directories.*

a. The Director of the Office of Management and Budget may require reports, from time to time, on the implementation of this Part.

b. The Office of Management and Budget will maintain and distribute to appropriate Federal agencies a directory of State, regional, and metropolitan clearinghouses.

c. The Office of Management and Budget will notify clearinghouses and Federal agencies of any excepted categories of projects under programs listed in Attachment D.

PART II: DIRECT FEDERAL DEVELOPMENT

1. *Purpose.* The purpose of this Part is to:

a. Provide State and local government with information on projected Federal development so as to facilitate coordination with State, regional, and local plans and programs.

b. Provide Federal agencies with information on the relationship of proposed direct Federal development projects and activities to State, regional, and local plans and programs; and to assure maximum feasible consistency of Federal developments with State, regional, and local plans and programs.

c. Provide Federal agencies with information on the possible impact on the environment of proposed Federal development.

2. *Coordination of direct Federal development projects with State, regional, and local development.*

a. Federal agencies having responsibility for the planning and construction of Federal buildings and installations or other Federal public works or development or for the acquisition, use, and disposal of Federal land and real property will establish procedures for:

(1) Consulting with Governors, regional, and metropolitan clearinghouses, and local elected officials at the earliest practicable stage in project or development planning on the relationship of any plan or project to the development plans and programs of the State, region, or localities in which the project is to be located.

(2) Assuring that any such Federal plan or project is consistent or compatible with State, regional, and local development plans and programs identified in the course of such consultations. Exceptions will be made only where there is clear justification.

(3) Providing State, metropolitan, regional, and local agencies which are authorized to develop and enforce environmental standards with adequate opportunity to review such Federal plans and projects pursuant to section 102(2) (C) of the National Environmental Policy Act of 1969. Any comments of such agencies will accompany the environmental impact statement submitted by the Federal agency.

3. *Use of clearinghouses.* The State, regional, and metropolitan planning and development clearinghouses established pursuant to Part I will be utilized to the greatest extent practicable to effectuate the requirements of this Part. Agencies are urged to establish early contact with clearinghouses to work out arrangements for carrying out the consultation and review required under this Part, including identification of types of projects considered appropriate for consultation and review.

PART III: STATE PLANS

1. *Purpose.* The purpose of this Part is to provide Federal agencies with information about the relationship of State plans required under various Federal programs to State comprehensive planning and to other State plans.

2. *Review of State plans.* To the extent not presently required by statute or administrative regulation, Federal agencies administering programs requiring by statute or regulation a State plan as a condition of assistance under such programs will require that the Governor be given the opportunity to comment on the relationship of such State plan to comprehensive and other State plans and programs. Governors will be afforded a period of forty-five days in which to make such comments, and any such comments will be transmitted with the plan.

3. *State plan.* A State plan under this Part is defined to include any required supporting reports or documentation that indicate the programs, projects, and activities for which Federal funds will be utilized.

PART IV: COORDINATION OF PLANNING IN MULTIJURISDICTIONAL AREAS

1. *Policies and objectives.* The purposes of this Part are:

    a. To encourage and facilitate State and local initiative and responsibility in developing organizational and procedural arrangements for coordinating comprehensive and functional planning activities.

    b. To eliminate overlap, duplication, and competition in State and local planning activities assisted or required under Federal programs and to encourage the most effective use of State and local resources available for development planning.

    c. To minimize inconsistency among Federal administrative and approval requirements placed on State, regional, and metropolitan development planning activities.

    d. To encourage the States to exercise leadership in delineating and establishing a system of planning and development districts or regions in each State, which can provide a consistent geographic base for the coordination of Federal, State and local development programs.

2. *Common or consistent planning and development districts or regions.* Prior to the designation or redesignation (or approval thereof) of any planning and development district or region under any Federal program, Federal agency procedures will provide a period of thirty days for the Governor(s) of the State(s) in which the district or region will be located to review the boundaries thereof and comment upon its relationship to planning and development districts or regions established by the State. Where the State has established such planning and development districts, the boundaries of designated areas will conform to them unless there is clear justification for not doing so. Where the State has not established planning and development districts or regions which provide a basis for evaluation of the boundaries of the area proposed for designation, major

units of general local government and Federal agencies administering related programs in such area will also be consulted prior to designation of the area to assure consistency with districts established under interlocal agreement and under related Federal programs.

3. *Common and consistent planning bases and coordination of related activities in multijurisdictional areas.* Each agency will develop checkpoint procedures and requirements for applications for planning and development assistance under appropriate programs to assure the fullest consistency and coordination with related planning and development being carried on under other Federal programs or under State and local programs in any multijurisdictional areas.

The checkpoint procedures will incorporate provisions covering the following points:

    a. Identification by the applicant of planning activities being carried on for related programs within the multijurisdictional area, including those covering a larger area within which such multijurisdictional area is located, subareas of the area, and areas overlapping the multijurisdictional area. Metropolitan or regional clearinghouses established under Part I of this Circular may assist in providing such identification.

    b. Evidence of explicit organizational or procedural arrangements that have been or are being established by the applicant to assure maximum coordination of planning for such related functions, programs, projects and activities within the multijurisdictional area. Such arrangements might include joint or common boards of directors or planning staffs, umbrella organizations, common referral or review procedures, information exchanges, etc.

    c. Evidence of cooperative arrangements that have been or are being made by the applicant respecting joint or common use of planning resources (funds, personnel, facilities, and services, etc.) among related programs within the area; and

    d. Evidence that planning being assisted will proceed from base data, statistics, and projections (social, economic, demographic, etc.) and assumptions that are common to or consistent with those being employed for planning related activities within the area.

4. *Joint funding.* Where it will enhance the quality, comprehensive scope, and coordination of planning in multijurisdictional areas, Federal agencies will, to the extent practicable, provide for joint funding of planning activities being carried on therein.

5. *Coordination of agency procedures and requirements.* With respect to the steps called for in paragraphs 2 and 3 of this Part, departments and agencies will develop for relevant programs appropriate draft procedures and requirements. Copies of such drafts will be furnished to the Director

of the Office of Management and Budget and to the heads of departments and agencies administering related programs. The Office, in consultation with the agencies, will review the draft procedures to assure the maximum obtainable consistency among them.

PART V: DEFINITIONS

Terms used in this Circular will have the following meanings:

1. *Federal agency*—any department, agency, or instrumentality in the executive branch of the Government and any wholly owned Government corporation.

2. *State*—any of the several States of the United States, the District of Columbia, Puerto Rico, any territory or possession of the United States, or any agency or instrumentality of a State, but does not include the governments of the political subdivisions of the State.

3. *Unit of general local government*—any city, county, town, parish, village, or other general purpose political subdivision of a State.

4. *Special purpose unit of local government*—any special district, public purpose corporation, or other strictly limited purpose political subdivision of a State, but shall not include a school district.

5. *Federal assistance, Federal financial assistance, Federal assistance programs, or federally assisted program*—programs that provide assistance through grant or contractual arrangements. They include technical assistance programs, or programs providing assistance in the form of loans, loan guarantees, or insurance. The term does not include any annual payment by the United States to the District of Columbia authorized by article VI of the District of Columbia Revenue Act of 1947 (D.C. Code sec. 47-2501a and 47-2501b).

6. *Comprehensive planning*, to the extent directly related to area needs or needs of a unit of general local government, includes the following:
    a. Preparation, as a guide for governmental policies and action, of general plans with respect to:
        (1) Pattern and intensity of land use,
        (2) Provision of public facilities (including transportation facilities) and other government services.
        (3) Effective development and utilization of human and natural resources.
    b. Preparation of long range physical and fiscal plans for such action.

c. Programming of capital improvements and other major expenditures, based on a determination of relative urgency, together with definitive financing plans for such expenditures in the earlier years of the program.

d. Coordination of all related plans and activities of the State and local governments and agencies concerned.

e. Preparation of regulatory and administrative measures in support of the foregoing.

7. *Metropolitan area*—a standard metropolitan statistical area as established by the Office of Management and Budget, subject, however, to such modifications and extensions as the Office of Management and Budget may determine to be appropriate for the purposes of section 204 of the Demonstration Cities and Metropolitan Development Act of 1966, and these Regulations.

8. *Areawide agency*—an official State or metropolitan or regional agency empowered under State or local laws or under an interstate compact or agreement to perform comprehensive planning in an area; an organization of the type referred to in section 701(g) of the Housing Act of 1954; or such other agency or instrumentality as may be designated by the Governor (or, in the case of metropolitan areas crossing State lines, any one or more of such agencies or instrumentalities as may be designated by the Governors of the States involved) to perform such planning.

9. *Planning and development clearinghouse* or *clearinghouse* includes:

a. An agency of the State Government designated by the Governor or by State law.

b. A nonmetropolitan regional comprehensive planning agency (herein referred to as "regional clearinghouse") designated by the Governor (or Governors in the case of regions extending into more than one State) or by State law.

c. A metropolitan areawide agency that has been recognized by the Office of Management and Budget as an appropriate agency to perform review functions under section 204 of the Demonstration Cities and Metropolitan Development Act of 1966.

10. *Multijurisdictional area*—any geographical area comprising, encompassing, or extending into more than one unit of general local government.

11. *Planning and development district or region*—a multijurisdictional area that has been formally designated or recognized as an appropriate area for planning under State law or Federal program requirements.

12. *Direct Federal development*—planning and construction of public

Office of Management Circular A-95, Revised 161

works, physical facilities, and installations or land and real property development (including the acquisition, use, and disposal of real property) undertaken by or for the use of the Federal Government or any of its agencies.

ATTACHMENT B
Circular No. A-95
Revised

SECTION 204 OF THE DEMONSTRATION CITIES AND METROPOLITAN DEVELOPMENT ACT OF 1966, as amended (80 Stat. 1263, 82 Stat. 208)

"Sec. 204. (a) All applications made after June 30, 1967 for Federal loans or grants to assist in carrying out open-space land projects or for planning or construction of hospitals, airports, libraries, water supply and distribution facilities, sewerage facilities and waste treatment works, highways, transportation facilities, law enforcement facilities, and water development and land conservation projects within any metropolitan area shall be submitted for review—

"(1) to any areawide agency which is designated to perform metropolitan or regional planning for the area within which the assistance is to be used, and which is, to the greatest practicable extent, composed of or responsible to the elected officials of a unit of areawide government or of the units of general local government within whose jurisdiction such agency is authorized to engage in such planning, and

"(2) if made by a special purpose unit of local government, to the unit or units of general local government with authority to operate in the area within which the project is to be located.

"(b)(1) Except as provided in paragraph (2) of this subsection, each application shall be accompanied (A) by the comments and recommendations with respect to the project involved by the areawide agency and governing bodies of the units of general local government to which the application has been submitted for review, and (B) by a statement by the applicant that such comments and recommendations have been considered prior to formal submission of the application. Such comments shall include information concerning the extent to which the project is consistent with comprehensive planning developed or in the process of development for the metropolitan area or the unit of general local government, as the case may be, and the extent to which such project contributes to the fulfillment of such planning. The comments and recommendations and the statement referred to in this paragraph shall, except in the case referred to in paragraph (2) of this subsection, be reviewed by the agency of the Federal Government to which such application is submitted for the sole purpose of assisting it in determining whether the application is in accordance with the provisions of Federal law which govern the making of the loans or grants.

"(2) An application for a Federal loan or grant need not be accompanied by the comments and recommendations and the statements referred to in paragraph (1) of this subsection, if the applicant certifies that a plan or description of the project, meeting the requirements of such rules and regulations as may be prescribed under subsection (c), or such application, has lain before an appropriate areawide agency or instrumentality or unit of general local government for a period of sixty days without comments or recommendations thereon being made by such agency or instrumentality.

"(3) The requirements of paragraphs (1) and (2) shall also apply to any amendment of the application which, in light of the purposes of this title, involves a major change in the project covered by the application prior to such amendment.

"(c) The Bureau of the Budget, or such other agency as may be designated by the President, is hereby authorized to prescribe such rules and regulations as are deemed appropriate for the effective administration of this section."

TITLE IV OF THE INTERGOVERNMENTAL COOPERATION ACT OF 1968 (82 Stat. 1103)

"TITLE IV—COORDINATED INTERGOVERNMENTAL POLICY AND ADMINISTRATION OF DEVELOPMENT ASSISTANCE PROGRAMS"

"DECLARATION OF DEVELOPMENT ASSISTANCE POLICY"

"Sec. 401. (a) The economic and social development of the Nation and the achievement of satisfactory levels of living depend upon the sound and orderly development of all areas, both urban and rural. Moreover, in a time of rapid urbanization, the sound and orderly development of urban communities depends to a large degree upon the social and economic health and the sound development of smaller communities and rural areas. The President shall, therefore, establish rules and regulations governing the formulation, evaluation, and review of Federal programs and projects having a significant impact on area and community development, including programs providing Federal assistance to the States and localities, to the end that they shall most effectively serve these basic objectives. Such rules and regulations shall provide for full consideration of the concurrent achievement of the following specific objectives and, to the extent authorized by law, reasoned choices shall be made between such objectives when they conflict:

"(1) Appropriate land uses for housing, commercial, industrial, governmental, institutional, and other purposes;

"(2) Wise development and conservation of natural resources, including land, water, minerals, wildlife, and others;

"(3) Balanced transportation systems, including highway, air, water, pedestrian, mass transit, and other modes for the movement of people and goods;

"(4) Adequate outdoor recreation and open space;

"(5) Protection of areas of unique natural beauty, historical and scientific interest;

"(6) Properly planned community facilities, including utilities for the supply of power, water, and communications, for the safe disposal of wastes, and for other purposes; and

"(7) Concern for high standards of design.

"(b) All viewpoints—national, regional, State and local—shall, to the extent possible, be fully considered and taken into account in planning Federal or federally assisted development programs and projects. State and local government objectives, together with the objectives of regional organizations shall be considered and evaluated within a framework of national public objectives, as expressed in Federal law, and available projections of future national conditions and needs of regions, States, and localities shall be considered in plan formulation, evaluation, and review.

"(c) To the maximum extent possible, consistent with national objectives, all Federal aid for development purposes shall be consistent with and further the objectives of State, regional, and local comprehensive planning. Consideration shall be given to all developmental aspects of our total national community, including but not limited to housing, transportation, economic development, natural and human resources development, community facilities, and the general improvement of living environments.

"(d) Each Federal department and agency administering a development assistance program shall, to the maximum extent practicable, consult with and seek advice from all other significantly affected Federal departments and agencies in an effort to assure fully coordinated programs.

"(e) Insofar as possible, systematic planning required by individual Federal programs (such as highway construction, urban renewal, and open space) shall be coordinated with and, to the extent authorized by law, made part of comprehensive local and areawide development planning."

"FAVORING UNITS OF GENERAL LOCAL GOVERNMENT"

"Sec. 402. Where Federal law provides that both special-purpose units of local government and units of general local government are eligible to receive loans or grants-in-aid, heads of Federal departments and agencies shall, in the absence of substantial reasons to the contrary, make such loans or grants-in-aid to units of general local government rather than to special-purpose units of local government."

"RULES AND REGULATIONS"

"Sec. 403. The Bureau of the Budget, or such other agency as may be designated by the President, is hereby authorized to prescribe such rules and regulations as are deemed appropriate for the effective administration of this title."

ATTACHMENT C
Circular No. A-95
Revised

SECTION 102 (2) (C) OF THE NATIONAL ENVIRONMENTAL POLICY ACT OF 1969 (83 Stat. 853)

"Sec. 102. The Congress authorizes and directs that, to the fullest extent possible: (1) the policies, regulations, and public laws of the United States shall be interpreted and administered in accordance with the policies set forth in this Act, and (2) all agencies of the Federal Government shall—

"(C) include in every recommendation or report on proposals

for legislation and other major Federal actions significantly affecting the quality of the human environment, a detailed statement by the responsible official on—

"(i) the environmental impact of the proposed action,

"(ii) any adverse environmental effects which cannot be avoided should the proposal be implemented,

"(iii) alternatives to the proposed action,

"(iv) the relationship between local short-term use of man's environment and the maintenance and enhancement of long-term productivity, and

"(v) any irreversible or irretrievable commitments of resources which would be involved in the proposed action should it be implemented.

"Prior to making any detailed statement, the responsible Federal official shall consult with and obtain the comments of any Federal agency which has jurisdiction by law or special expertise with respect to any environmental impact involved. Copies of such statement and the comments and views of the appropriate Federal, State, and local agencies, which are authorized to develop and enforce environmental standards, shall be made available to the President, the Council on Environmental Quality and to the public as provided by section 552 of Title 5, United States Code, and shall accompany the proposal through the existing agency review processes; . . . ."

APPENDIX 4

National List
of
State Clearinghouses

ALABAMA
Alabama Development Office
State Office Building
Montgomery, Alabama 36104

ALASKA
Office of the Governor—
State of Alaska
Division of Planning & Research
Pouch AD
Juneau, Alaska 99801

ARIZONA
Dept. of Economic Planning & Development
State of Arizona
Phoenix, Arizona 85007

ARKANSAS
Arkansas Planning Commission
Room 300, Game & Fish Commission Building
Little Rock, Arkansas 72201

CALIFORNIA
Office of the Lieutenant Governor
Office of Intergovernmental Management
Sacramento, California 95814

COLORADO
State Planning Coordinator
Governor's Office
Denver, Colorado 80203

CONNECTICUT
Office of State Planning
Dept. of Finance and Control
340 Capitol Avenue
Hartford, Connecticut 06115

DELAWARE
Delaware State Planning Office
Tomas Collins Building
530 South Dupont Highway
Dover, Delaware 19901

DISTRICT OF COLUMBIA
Div. of Budget & Program Analysis
Office of Budget & Executive Management
Government of the District of Columbia
Washington, D. C. 20004

FLORIDA
Dept. of Administration
Tallahassee, Florida 32304

GEORGIA
Bureau of State Planning & Community Affairs
Room 611
270 Washington Street, S.W.
Atlanta, Georgia 30303

HAWAII
Dept. of Planning & Economic Development
P. O. Box 2359
Honolulu, Hawaii 96804

IDAHO
Division of State Planning
State Capitol
Boise, Idaho 83702

ILLINOIS
State Clearinghouse
Office of the Governor
205 State House
Springfield, Illinois 62706

INDIANA
Office of the Governor
206 State House
Indianapolis, Indiana 46204

IOWA
Office of Planning and Programming
Des Moines, Iowa 50319

KANSAS
Budget Division

National List of State Clearinghouses

Dept. of Administration
State Capitol Building
Topeka, Kansas 66612

KENTUCKY
Kentucky Program Development Office
Room 157, Capitol Building
Frankfort, Kentucky 40601

LOUISIANA
Commission on Intergovernmental Relations
P. O. Box 44316
Baton Rouge, Louisiana 70804

MAINE
State Planning Office
Executive Department—State of Maine
189 State Street
Augusta, Maine 04330

MARYLAND
Dept. of State Planning
301 West Preston Street
Baltimore, Maryland 21201

MASSACHUSETTS
Office of Planning & Programming Coordination
209 Leverett Saltonstall Building
100 Cambridge Street
Boston, Massachusetts 02202

MICHIGAN
Office of Planning Coordination
Executive Office of the Governor
Lewis Cass Building
Lansing, Michigan 48913

MINNESOTA
Minnesota State Planning Agency
Suite 603
550 Cedar Street
St. Paul, Minnesota 55101

MISSISSIPPI
Coordinator Federal-State Programs
Office of the Governor
510 Lamar Life Building
Jackson, Mississippi 39201

MISSOURI
Missouri Dept. of Community Affairs
500 Jefferson Building
Jefferson City, Missouri 65101

MONTANA
State Dept. of Planning & Economic Development
Old Governor's Mansion
Helena, Montana 59601

NEBRASKA
State Office of Planning & Programming
Box 94601
Lincoln, Nebraska 68509

NEVADA
Chief, Budget Division
Dept. of Administration
Carson City, Nevada 89701

NEW HAMPSHIRE
Special Assistant for Planning
Office of the Governor
State House
Concord, New Hampshire 03301

NEW JERSEY
Div. of State & Regional Planning
Dept. of Community Affairs
P. O. Box 1978
Trenton, New Jersey 08625

NEW MEXICO
State Planning Office
Santa Fe, New Mexico 87501

NEW YORK
New York State Office of Planning
  Coordination
488 Broadway
Albany, New York 12207

NORTH CAROLINA
Planning Coordinator
Clearinghouse and Information
  Center
P. O. Box 1351
Raleigh, North Carolina 27602

NORTH DAKOTA
North Dakota State Planning
  Agency
Bismarck, North Dakota 58501

OHIO
Office of the Governor
Planning & Development Clearing-
  house
Box 1001
Columbus, Ohio 43215

OKLAHOMA
Office of the Director of State
  Finance
Office of the Governor
Oklahoma City, Oklahoma 73105

OREGON
Office of the Governor
Executive Department
Salem, Oregon 97310

PENNSYLVANIA
Pennsylvania State Planning Board
503 Finance Building
Harrisburg, Pennsylvania 17120

PUERTO RICO
Planning Board
1507 Ponce de Leon Avenue
Cond. Ponce de Leon
Box 9447
Santurce, Puerto Rico 00908

RHODE ISLAND
Statewide Planning Program
Room 123A
State House
Providence, Rhode Island 02903

SOUTH CAROLINA
Office of the Governor
State Planning & Grants Division
915 Main Street
Columbia, South Carolina 29201

SOUTH DAKOTA
State Planning Agency & the Office
  of the Budget
Pierre, South Dakota 57501

TENNESSEE
Office of Urban and Federal Affairs
321 Seventh Avenue, North
Nashville, Tennessee 37219

TEXAS
Division of Planning Coordination
Office of the Governor
Drawer P, Capitol Station
Austin, Texas 78711

UTAH
Utah State Planning Coordinator
Office of the Governor
Salt Lake City, Utah 84114

VERMONT
Planning & Community Services
  Agency
State Office Building
Montpelier, Vermont 05602

VIRGINIA
Virginia Division of Planning &
  Community Affairs
1010 James Madison Building
Richmond, Virginia 23219

WASHINGTON

Office of the Governor
State Planning Division & Community Assistance Division
100 Insurance Building
Olympia, Washington 98501

WEST VIRGINIA

Grant Information Department
Office of Federal-State Relations
1703 Washington Street, East
Charleston, West Virginia 25311

WISCONSIN

State Planning Bureau
Dept. of Administration
1 West Wilson Street
State Office Building
Madison, Wisconsin 53701

WYOMING

State Planning Coordinator
Office of the Governor
Capitol Building
Cheyenne, Wyoming 82001

APPENDIX
5

Section 4(f)

"Section 4(f)" permits the Secretary of Transportation to approve a program or project which requires the use of publicly owned land from a park, recreation area, or wildlife and waterfowl refuge of national, State, or local significance as determined by the Federal, State or local officials having jurisdiction thereof: or land from an historic site of national, State or local significance as so determined by such officials (hereafter "Section 4(f) land") only if:

(1) there is no feasible and prudent alternative to the use of such land, and
(2) such program includes all possible planning to minimize harm to the Section 4(f) land resulting from such use.

The application, according to Federal Highway Administration's (FHWA) PPM 90-1 should contain:

a. The description of the project shall include information about the Section 4(f) land in sufficient detail to permit those not acquainted with the project to have an understanding of the relationship between the highway and park and the extent of the impact, such as:

(1) Size (acres or square feet) and location (maps or other exhibits such as photographs, slides, sketches, etc. as appropriate).
(2) Type (recreation, historic, etc.).
(3) Available activities (fishing, swimming, golf, etc.).
(4) Facilities existing and planned (description and location of ball diamonds, tennis courts, etc.).
(5) Usage (approximate number of users for each activity if such figures are available).
(6) Patronage (local, regional, and national).
(7) Relationship to other lands similarly used in the vicinity.
(8) Access (both pedestrian and vehicular).
(9) Ownership (city, county, State, etc.).
(10) If applicable, deed restrictions or reversionary clauses.
(11) The determination of significance by the Federal, State, or local officials having jurisdiction of the Section 4(f) land.
(12) Unusual characteristics of the Section 4(f) land (flooding problems, terrain conditions, or other features that either reduce or enhance the value of portions of the area).
(13) Consistency of location, type of activity, and use of the Section 4(f) land with community goals, objectives, and land use planning.
(14) If applicable, prior use of State or Federal funds for acquisition or development of the Section 4(f) land.

b. A description of the manner in which the highway will affect the

Section 4(f) land (include within paragraph 2c of this Appendix such as):

(1) The location and amount of land (acres or square feet) to be used by the highway.
(2) A detailed map or drawing of sufficient scale to discern the essential elements of the highway/Section 4(f) land involvement.
(3) The facilities affected.
(4) The probable increase or decrease in physical effects on the Section 4(f) land users (noise, fumes, etc.).
(5) The effect upon pedestrian and vehicular access to the Section 4(f) land.

c. A specific statement (with supporting reasons) that there is no feasible and prudent alternative.

d. Information to demonstrate that all possible planning to minimize harm is, or will be, included in the highway proposal. Such information should include:

(1) The agency responsible for furnishing the highway right-of-way.
(2) Provisions for compensating or replacing the Section 4(f) land and improvements thereon, including the status of any agreements. (Include agreed upon compensation, replacement acreages, and type of land, etc., when known.)
(3) Highway design features developed to enhance the Section 4(f) land or to lessen or eliminate adverse effects (improving or restoring existing pedestrian or vehicular access, landscaping, esthetic treatment, etc.).
(4) Coordination of highway construction to permit orderly transition and continual usage of Section 4(f) land facilities (new facilities constructed and available for use prior to demolishing existing facilities, moving of facilities during off-season, etc.).

E. Evidence that the provisions of 16 U.S.C. 470(f) (Section 106 of the Historic Preservation Act of 1966) have been satisfied when National Register Properties are involved.

APPENDIX
6

Mathematical Evaluation System

The EIS process is an attempt to assess environmental impacts of various alternatives established to solve a given problem. This includes the "trade-offs" of go or no-go as well as the effects of each of the options set up to address the given problem.

The major problem is to place the "trade-offs" and various impacts into some perspective, e.g., weighted values. Some of the factors can be objectively quantified, as, for instance, construction costs. However, more often the values have to be weighted on a qualitative, thus subjective basis.

To avoid the more direct effect of a single number placed on a particular item or factor, this system was designed so that several points of view can be included in the weighting. In addition, the opinion of each contributor of the before-and-after value of the specific item is measured against several basic yardsticks.

It is hoped that the weighting will be less directly subjective and given a broader basis upon which to establish the item value and final totals.

Within this system the features of primary interest are those that will be affected, whether in the area of take or not, and the associated effects, both positive and negative, of the project on the environment. The goal is to find the procedure that sacrifices the least and nets the most for society, that is, a "trade-off," where the plus factors (benefits) outweigh the minus factors (losses). Part of the evaluation process is a series of judgments that weigh each item and evaluate its importance. Since these values change in different relationships, weighting has many facets to be considered.

In the author's opinion, there are four major classifications that include the significant conditions associated with every type of project. They are *socio-economics, aesthetics, natural resources,* and *biotics.* The total of these relates the local ecological system to man. Under these general categories, it is possible to associate every specific feature to a proposed action. When these factors are weighted and totaled, it should be possible to determine whether or not the exchange will be advantageous. To accomplish this in as objective a manner as possible, the system can be approached from a mathematical viewpoint. The first requirement is a weighting system which has been arbitrarily established as a scale from 0 to 9. Zero would then represent minimum value and nine, the greatest. The evaluators, as they place a value on each item, consider both the item and its relationship to the conditions of the area. The second requirement is a format whereby factors can be added or subtracted on a before-and-after relationship. The totals should indicate whether the final outcome is to be favorable or not for the environment, or whether damage to the environment can be tolerated in light of other benefits to be gained. Several features will be examined by persons with various interests.

Mathematical Evaluation System

The following system was devised with these requirements in mind. An example will be presented later in the section.

INDIVIDUAL ITEM EVALUATION SHEET

Associate: _____

Category: _____ Item Number: _____

Feature: _____

Description:

General Conditions:

Reason for Destruction, Alteration, or Pollution:

Effect of Destruction, Alteration, or Pollution:

Effect		Degree			
Positive	Negative	Large	Moderate	Small	None

Weighted Value—Existing:

Socio-Economics   Aesthetics   Natural Resources   Biotics

Weighted Value—Projected after Destruction or Alteration:

Socio-Economics   Aesthetics   Natural Resources   Biotics

The Individual Item Sheet contains the item number for identification, a brief description, and its general relationship. Both negative and positive effects will be addressed.

An item is weighted in each of the four classifications, according to its value before and also after change; then the net gain or loss is noted. If there is more than one evaluation for any item, the weights are averaged. Features that are adversely affected will of course indicate a net loss. Some features that are being added, or those changed beneficially will indicate a net gain.

The sum of the four classifications offers one way to assess the environmental impact of any given project. The gains and losses are also totaled to determine the net impact on the system. The completed Tabulation Form briefly demonstrates a weighting exercise. For instance, as shown in the form, a new reservoir area was evaluated by Dr. Jones, a terrestrial ecologist. Before construction, he allowed it a value of 2 on the 0–9 scale as related to socio-economics. After construction he evaluated the area at 7. Thus construction resulted in a net gain of 5 in the socio-economics column.

In the matter of aesthetics, he rated the area as 3 before construction and 6 afterward, indicating an increase of 3.

Thus there was an increase of 2 regarding natural resources and 1 regarding biotics.

The total of the weighting before construction shown in the "Total Impact Column" adds up to 9. After construction the value rises to 20 as shown on the second line. Thus there would be a positive impact on the environment of 11 after construction of the proposed reservoir.

Another assessor, however, evaluated the same area from the viewpoint of an aquatic biologist and came up with different figures, as shown in item 1-B. These are averaged and used in the final tally. Item 2, a woodcock breeding area was evaluated by Dr. Brown, a wildlife manager. He indicates a net loss of 2 regarding wildlife in the area.

When all the figures are determined and the net gain or loss is calculated, the sum of these figures will indicate the net environmental effect of the project. In addition, percentages can be determined for every column. The number and type of factors involved in this system will be determined by the actual circumstances.

Each item that may be disturbed should be included and weighted, as any omission will prevent the final figures from giving an accurate evaluation of the specific project. This must include intangible as well as tangible factors, to be complete.

Each evaluator will no doubt exercise bias in relation to his own area of expertise, however, so being aware of the total picture could allow for a broader base of judgment.

As previously mentioned, all weighting systems contain subjective input that limits their value. The author has attempted to dilute this effect by averaging the values of several inputs on the same item and simultaneously measuring those items against four basic scales.

Nevertheless, subjectivity cannot be totally eliminated. A useful means of controlling this problem is to establish proper understanding among various specialists, so that their evaluations are consistent and compatible.

A more involved example follows.

SIMULATED MATHEMATICAL EVALUATION PROBLEM

Proposed: A new Pump Storage Power Generating Facility.

Problem: To determine suitability of a specific site for proposed projects.

PROJECT CONSIDERATIONS

The area under assessment is in a relatively undeveloped location in mountainous terrain. Construction of the pump storage system requires that two sites be inundated. One reservoir will cover approximately 400 acrees and the other, 4000 acres. The owner already has possession of the acreage under discussion; however, he is concerned with objections to the project that have been instituted by interested parties. The purpose of the evaluation is to determine what effect the project will have on the area, and the net result to society.

The team considering this project was composed of a general ecologist, an aquatic biologist, a wildlife manager, and a socio-economist.

An analysis of the area indicated these existing conditions:

(a) An existing stream was highly eutrophic because of a large population but carried little or no game fish population.

(b) The area was covered by secondary scrub vegetation. This is the result of heavy timbering activities earlier in the century.

(c) The swamp was not distinctive or valuable in any sense and was not a major breeding ground for wild fowl.

(d) The deer population was limited, and the inundation would not force a serious migration. In fact, conditions would be generally improved for most wild game.

(e) The area was economically depressed. Employment opportunities were limited and the tax base low. There were a limited number of

TABULATION FORM

Item	Associate and Specialty	Time Element As Related to Conservation	Weighted Value to Socio-Economics ←0–9→	Weighted Value to Asthetics ←0–9→	Weighted Value to Natural Resources ←0–9→	Weighted Value to Biotics ←0–9→	Total Environmental Impact ←0–36→
1a. New Reservoir	Jones General Ecologist	Before	2	3	2	2	9
		After	7	6	4	3	20
		Net Loss/Gain	+5	+3	+2	+1	+11
1b. New Reservoir	Smith Aquatic Biologist	Before	2	3	1	1	7
		After	5	2	3	5	15
		Net Loss/Gain	+3	−1	+2	+4	+8
1. Arrived at by Total ÷ No. of Associates		Consensus Item 1	+4	+1	+2	+2.5	+9.5
2. Woodcock Breeding Area	Brown Wildlife Management	Before	2	3	1	2	8
		After	0	4	1	1	6
		Net Loss/Gain	−2	+1	0	−1	−2

182

TABULATION FORM (CONTINUED)

Item	Associate and Specialty	Time Element As Related to Conservation	←0–9→ Weighted Value to Socio-Economics	←0–9→ Weighted Value to Aesthetics	←0–9→ Weighted Value to Natural Resources	←0–9→ Weighted Value to Biotics	←0–36→ Total Environmental Impact
3. Tax Base	Black Socio-Economics	Before	3	0	0	0	3
		After	8	0	0	0	8
		Net Loss/Gain	+5	0	0	0	+5
4. Local Plant Life	Robert Botanist	Before	1	2	0	3	6
		After	1	1	0	4	6
		Net Loss/Gain	0	−1	0	+1	0
5. New Recreation Area	Schultz Landscape Architect	Before	2	2	0	0	4
		After	5	4	0	0	9
		Net Loss/Gain	+3	+2	0	0	+5
Total Loss/Gain by Classification			+10	+3	+2	+2.5	+17.5

hunters coming into the area during the hunting season, but this did not amount to a major influx.

(f) There was a large state park nearby, also an area disturbed by strip mining, and a fossil fueled power station.

The small reservoir is not in contention, as it will cover an area of little importance, is not of sufficient size to be objectionable, and will be well screened when completed. The larger area has come under attack on the basis that a significant swamp will be destroyed, and a prime game fowl hunting area will be lost. As the site is located in Appalachia, the local economic conditions are relatively poor, and those living in the area are generally enthusiastic about the project.

Within the 4,000 acres to be flooded, the following features, slated for destruction, were claimed to have considerable value by those objecting to the construction:

(a) Swamp area;
(b) A hunting area, with special emphasis on game fowl;
(c) A wildlife area that supported a deer population;
(d) The natural condition or "wilderness" environment of the area.

However, when the facility was completed, some positive results would include:

1. A 4,000 acre lake covering what was an ecologically distressed area. This reservoir, in spite of a 2-foot draw down or lowering of the water level when there was a need for power, could support a sizable fish population. Dams across small fingers of the lake could insure natural breeding areas.

2. The value of the installation would considerably increase the local tax base.

3. Employment would rise during the three-year construction period and, when the project was completed, several permanent positions would be available for local people.

4. The lake would become a recreational facility to include such activities as fishing, bathing, and boating. Camping grounds and picnic facilities could be added, if they were properly managed, to avoid destruction of the area's natural beauty.

5. The local community would benefit from the money brought into the area by tourists, hunters, fishermen, and other visitors to the recreation area.

6. With proper landscaping, the area would be aesthetic, and support reforestation programs.

The evaluation team viewed the main features of the project as follows:

Swamp area. This area, though not unique, has some aesthetic value, which will be lost when the area is inundated. The swamp has no relationship to socio-economics or natural resources.

Wildlife area. This area, which will be partially destroyed, brings hunters and, thus, income to the local community. If inundation occurs, this will be somewhat reduced, but not totally. Therefore, there will be a loss as specifically related to socio-economics. There is some aesthetic value, although it is not a spectacular area. However, any loss from this standpoint is negligible as the inundation creates a shore line of at least equal attraction. There will be a loss of biotic factors, as there will be less acreage available to support both flora and fauna.

Bluegill stream. This stream, made highly eutrophic by beavers, has no socio-economic value. The presence of the beavers while of some aesthetic value, is overbalanced by the pollution they stimulate, hence the stream has a negative aesthetic value. With the inundation, this area will totally disappear and this results in a net, but minor, loss.

Recreational facilities. As previously mentioned, the area currently attracts a limited number of hunters. With the filling of the lake, this number may be reduced slightly, but fishing interest will sharply increase. Plans include special recreational areas and facilities that will utilize the water potential. This influx will require lodging and eating facilities and other tourist necessities. Therefore, there will be a decisive gain in the impact on socio-economic factors. This same reasoning applies to aesthetics and, again, there will be a marked increase. These facilities will have no effect on natural resources and only limited effects on biotic factors.

Reservoir. If constructed, it will have a definite socio-economic and aesthetic impact. As fish will multiply in this new ecological system, it has an impact on biotic factors too.

Note that this item is evaluated twice. This is because two different specialists felt that it was of concern to their area of interest. Thus, each independently weighted the items from his viewpoint and as they related, in this case, to general ecology and aquatic biology. The final weighting of this item was obtained by averaging the two evaluations. When an item is to be evaluated more than once, the average of these weights should be used in the totaling of the figures.

Tax base. With the event of construction, the local tax base will increase because of the increased value of the property and the pump station itself. Actually, a spin-off of the project and the ensuing tourist accommodations and miscellaneous supporting businesses would further raise this base. This item has no effect on the other classifications.

EVALUATION SHEET 1

Project: Pump Storage Station
Location: Brunswick, Pa.

Item	Associate and Specialty	Time Element As Related to Construction	Weighted Value to Socio-Economics ←0—9→	Weighted Value to Asthetics ←0—9→	Weighted Value to Natural Resources ←0—9→	Weighted Value to Biotics ←0—9→	Total Environmental Impact ←0—36→
1. Swamp Area	E. F. Smith General Ecologist	Before	0	1	0	2	3
		After	0	0	0	0	0
		Net Loss/Gain	0	−1	0	−2	−3
2. Wildlife Area	B. C. Johns Wildlife Manager	Before	2	1	0	3	6
		After	1	1	0	2	4
		Net Loss/Gain	−1	0	0	−1	−2
3. Tax Base	A. R. Roger Socio-Economist	Before	3	0	0	0	3
		After	5	0	0	0	5
		Net Loss/Gain	+2	0	0	0	+2
4. Bluegill Stream	A. J. Fish Aquatic Biologist	Before	0	1	0	1	2
		After	0	0	0	0	0
		Net Loss/Gain	0	−1	0	−1	−2

EVALUATION SHEET 1 (CONTINUED)

	Associate and Specialty	Time Element As Related to Construction	←0–9→ Weighted Value to Socio-Economics	←0–9→ Weighted Value to Asthetics	←0–9→ Weighted Value to Natural Resources	←0–9→ Weighted Value to Biotics	←0–36→ Total Environmental Impact
5. Recreational Facilities	A. R. Roger Socio-Economist	Before	2	1	0	0	3
		After	5	4	0	0	9
		Net Loss/Gain	+3	+3	0	0	+6
6a. Reservoir	E. F. Smith General Ecologist	Before	0	0	0	0	0
		After	4	2	0	3	9
		Net Loss/Gain	+4	+2	0	+3	+9
6b. Reservoir	A. J. Fish Aquatic Biologist	Before	0	0	0	0	0
		After	2	1	0	4	7
		Net Loss/Gain	+2	+1	0	+4	+7
6c. Consensus on Reservoir		Net Loss/Gain	6÷2=+3	3÷2=+1.5	0	7÷2=+3.5	+8
7. Local Employment	A. R. Roger Socio-Economist	Before	2	0	0	0	2
		After	3	0	0	0	3
		Net Loss/Gain	+1	0	0	0	+1
Total Gain/Loss by Classification			+8.0	+1.5	0	−0.5	+10

187

Local employment. There would be a sharp increase in available jobs with the start of actual construction. This would extend through the time required to complete the project. However, when completed, and the construction employment halted, there would still be positions open to permanently staff the operation. Thus, there would be a permanent impact on socio-economics. This has no effect on the other classifications.

FINDINGS (SEE EVALUATION SHEETS)

The totals indicate a gain in every classification except biotics where the loss would amount to 3. The weights total up to a +10 out of a possible 324. This indicates that the consensus of the team was that the project is ecologically sound.

APPENDIX 7

Sample Draft Environmental Statement

SAMPLE

DRAFT ENVIRONMENTAL ASSESSMENT

OF

JENNERT MARINE TERMINAL

BRUNSWICK, NEW JERSEY

CONTENTS

	PAGE
LIST OF FIGURES	A- 3
LIST OF TABLES	A- 4
SUMMARY	A- 4
1. INTRODUCTION	
1.1 Project Surroundings	A- 5
1.2 Background	A- 7
2. ENVIRONMENTAL INVENTORY	
2.1 Climate	A- 8
2.2 Air Quality	A- 9
2.3 Acoustics	A- 9
2.4 Water Quality	A- 11
2.5 Geology	A- 18
2.6 Terrestrial Ecology	A- 19
2.7 Aquatic Biology	A- 20
2.8 Socio-Economics	A- 23
2.9 Historical/Archeological Sites	A- 23
3. PROPOSED ACTION AND ALTERNATIVES	
3.1 Project Description	A- 23
3.2 Engineering Data	A- 24
3.3 Alternatives	A- 30
4. ENVIRONMENTAL IMPACTS	
4.1 Climate	A- 30

Sample Draft Environmental Statement A-3

    4.2 Air Quality   A- 30
    4.3 Acoustics   A- 32
    4.4 Water Quality   A- 32
    4.5 Geology   A- 34
    4.6 Terrestrial Ecology   A- 34
    4.7 Aquatic Biology   A- 35
    4.8 Socio-Economics   A- 36
    4.9 Historical/Archeological Sites   A- 38
    4.10 Summary Effects   A- 38

5. UNAVOIDABLE ADVERSE EFFECTS WHICH CANNOT BE AVOIDED SHOULD THE PROJECT BE IMPLEMENTED   A- 39

6. MEASURE UNDER CONSIDERATION TO MINIMIZE UNAVOIDABLE ENVIRONMENTAL EFFECTS   A- 39

7. THE RELATIONSHIP BETWEEN LOCAL SHORT TERM USES OF MAN'S ENVIRONMENT AND THE MAINTENANCE AND ENHANCEMENT OF LONG TERM PRODUCTIVITY   A- 40

8. ANY IRREVERSIBLE AND IRRETRIEVABLE COMMITMENT OF RESOURCES WHICH WOULD BE INVOLVED IN THE PROPOSED ACTION, SHOULD IT BE IMPLEMENTED   A- 40

LIST OF FIGURES

			PAGE
1.	MAP	Surrounding Area of Brunswick	A- 6
2.	MAP	General Location and Well Sites	A- 16
3.	FIGURE	Cross Sections of Localized Typical Geology	A- 19
4.	PLAN	Three Proposed Phases of Construction	A- 25
5.	PLAN	Detail on Phases I and II	A- 26
6.	FIGURE	Construction Sections	A- 27
7.	FIGURE	Construction Sections	A- 28

LIST OF TABLES

			PAGE
1.	TABLE	Average Pollution Levels for Brunswick Area, 1971	A- 10
2.	TABLE	Federal Standard for Air Quality	A- 11
3.	TABLE	Water Quality Characteristics of Foster Creek at Front Avenue Station	A- 12
4.	TABLE	Water Quality Characteristics of Columbiana River at Stations 37, 14, and 13	A- 13
5.	TABLE	Permissible Water Quality Levels for Potable Water	A- 14
6.	TABLE	Quality of Water from Several Wells in the Area	A- 15
7.	TABLE	Ground Water Levels	A- 17
8.	TABLE	Analysis of Typical Incinerator Residue Leachate	A- 18
9.	LISTING	Species of Aquatic Life of Columbiana River near Chesville	A- 21
10.	LISTING	Species of Aquatic Life of Foster Creek	A- 22
11.	TABLE	Projected Air Polluting Emissions Generated by Project	A- 31
12.	TABLE	Projected Revenues Generated by Jennert Marine Terminal	A- 37

SUMMARY

Jennert Marine Terminal

(X) Draft Environmental Statement ( ) Final Environmental Statement

Responsible Office: (Complete Address & Phone No. of District Engineer)

1. Name of Action: ( ) Administrative   (X) Legislative

Description of Project: Construction of a Marine Terminal on the Columbiana in Brunswick, New Jersey in three phases. The project requires the rechannelization of Foster Creek and an encroachment into the Columbiana River not to exceed 395 feet. Embankments will be constructed and the area behind them filled with acceptable material. Considerable dredging activity will occur during construction.

Environmental Impact: The Project will result in additional facilities and revenue for the city of Brunswick making use of heretofore unproductive city owned land. The construction will have some temporary impacts on the Columbiana River and some permanent effects on Foster Creek.

Adverse Environmental Effects: Temporary turbidity and its effects on aquatic biota will result from intermittent dredging activity. Foster Creek will lose its natural meandering course and the ecosystem will be altered. Residents of the Golden Age Home and the Vo-Tech School will be subjected to some noise levels.

Alternatives: Eliminate phases two and/or three to reduce extent of impacts; do not construct project (no construction).

DRAFT ENVIRONMENTAL ASSESSMENT

JENNERT MARINE TERMINAL

BRUNSWICK, NEW JERSEY

1. INTRODUCTION

1.1 Project Surroundings

The proposed project, sponsored by the Brunswick Port Authority, will be located in Brunswick, New Jersey (Figure 1), the thirteenth largest metropolitan area in the United States. Brunswick's population and economic bases are relatively healthy and fall in line with today's national pictures. There is no trend towards a reduced census; however, like most large urban areas, it has many critical social and economic problems requiring attention. Employment figures are within range of national averages. The metropolitan area is highly industrialized, with no significant amount of agricultural activity.

Uniport is the term which encompasses all Ports of Brunswick on both sides of the Columbiana River from Monthouth, Pa. to Jonesville,

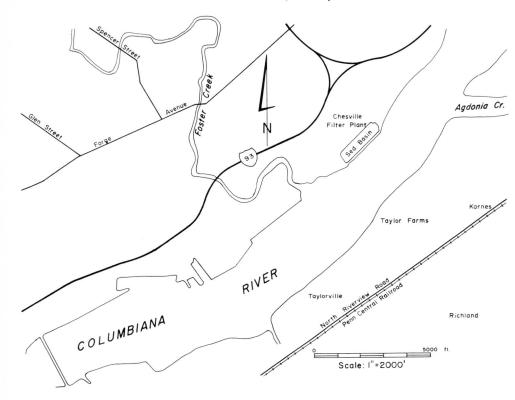

Figure 1 Surrounding Area of Brunswick.

Del. The unit represents a very large fresh water port and is one of the country's largest industrial ports. It is further, the world's third largest industrial port.

The region of which Brunswick is the hub is a heavily industrialized area and also a key transportation center.

Uniport serves over 100 steamship lines and last year handled 1.7 billion dollars worth of cargo, representing 100.2 million dollars in U. S. Government customs' receipts. Most of this material was in bulk form rather than general cargo. Handling general cargo is more profitable than bulk cargo. Jennert Terminal, as a new facility with the latest equipment, purportedly will provide for greater capacity and more efficient operation.

Specifically, the project is in northeast Brunswick on the banks of the Columbiana River. Directly upstream is the city's Chesville Filter Plant. The shoreline beyond that point, generally city owned, is not industrialized. The area is targeted for future recreational use.

Directly downstream is the Smith Forge Company Wharf. For

Sample Draft Environmental Statement

A-7

several miles below the wharf, the entire riverfront is highly industrialized.

Interstate 101 and State Road, approximately a half mile northwest of the site, separate the project from any residential areas.

The immediate vicinity includes the Golden Age Nursing Home, an institution devoted to caring for the aged, the Brunswick Vocational School, and the Brunswick County Jail. There are no registered historical or archeological landmarks in the vicinity.

Foster Creek borders a portion of the area and traverses another portion as it reaches its confluence with the Columbiana River. Except for the small section of creek and wooded area, the location is not aesthetically sensitive. In fact, a landfill area for incinerator residue is included in the study perimeters, making the overall location somewhat bleak.

The Columbiana River at this point is approximately six-tenths of a mile wide. The navigational channel is ±800 feet from the edge of the existing shoreline and ±265 feet from the deepest proposed encroachment into the river. It has been dredged to a depth of approximately 40 feet.

From the study point, the river is navigable for 25 miles upstream, and 98 miles downstream to the ocean. Ships up to 100,000 dead-weight tons can be accommodated to Brunswick. The annual net value volume of commerce served by the river in 1972 was 102,653,500 short tons.[1]

Area topography is relatively flat with a maximum elevation differential of 10 feet within the project area. The landfill, which produces the largest variation in ground elevation, is a temporary feature subject to grading changes.

Soil erosion and resulting sedimentation is uncommon, as most of the area is extremely flat, and a bulkhead has been constructed along the shoreline. Foster Creek does contribute some sediment load to the Columbiana; however, the meanders reduce this volume. The landfill on the site is a dumping area for city incinerators and presents a leaching problem which will be discussed later in the study.

City institutions are served by a system of storm sewers which empty into the Columbiana River and Foster Creek.

1.2 Background

The Brunswick Port Authority was chartered in April 1950:

> To promote waterborne commerce of the Port of Brunswick.

---

[1] Waterborne Commerce of the United States, Part I, Waterways and Harbors—Mid Continent, Department of the Army Corps of Engineers, 1971.

To acquire by lease or otherwise, maintain and modernize the Port's existing facilities for the handling of cargo.

To design, construct, lease or otherwise acquire and maintain and modernize new facilities in a comprehensive program for the prompt and full development of this Port's cargo handling potential, thereby stimulating industrial development, expanding employment opportunities and increasing tax revenues to the State of New Jersey, the city of Brunswick, and all other areas affected by the prosperity of this Port.

An independent authority, it consists of government officers and community representatives. Its primary function is the construction, management, and operation of Brunswick's public port facilities.

The firm was created to rectify adverse conditions pinpointed by a survey of coastal port facilities. It was determined that competing ports had activated terminal construction programs which attracted sufficient handling from 13.9% to 9.2% over a thirty-year period. This drop represented a loss of employment possibilities, revenue, and tax dollars for the city. In order to reverse this losing trend, a major immediate-action program was established. To expedite the plan, a public corporation was created to overcome existing city-state procedures.

It was determined that fifteen to twenty-one new general cargo berths would be needed over the next ten years in addition to the three-berth Rogers Avenue Marine Terminal.

Since that time Wilson Marine Terminal (seven berths), has been completed and placed in operation. The proposed Jennert Terminal is a new phase of the ongoing program and has been approved by the board and city officials.

All land essential to the project is cityowned. Negotiations are currently under way to purchase the adjoining Smith Forge Company so that its acreage can be included as part of the Jennert facility.

2. ENVIRONMENTAL INVENTORY

2.1 Climate [2]

Freedom Bay has some moderating effect on the temperature conditions in the Brunswick area. Extended periods of cold weather are comparatively rare and sub-zero readings have only been reported 24 times

---

[2] *Source:* U.S. Department of Commerce, National Oceanic and Atmospheric Administration, Environmental Data Service.

since official records have been maintained. Periods of extremely low temperatures are rare and generally don't last more than three or four days. The average annual temperature is 46.3°.

Maritime air, with its high humidity, causes discomfort during summer months; however, with the exception of the Mid City Airport, located almost on the river, heavy fog seldom occurs over a large section of the city. The average occurrence of fog is no more than ten times per year.

Precipitation is generally evenly distributed throughout the year, with maximum fall during the late summer months. The average total annual precipitation is 40.97 inches.

The prevailing wind direction in the summer months is from the southwest and during the winter, northwesterly. The annual prevailing direction is from the south-southwest. The mean wind speed is 9.6 miles per hour.

Flood stages on the Columbiana are the result of abnormally high tides that occur due to water backing up under the influence of strong south or southeast winds.

2.2 Air Quality

Monitoring at a station near the Jennert established the following pollution levels for the year 1971.[3] (Table 1)

Pollutant levels are reasonably high as would be expected in a large metropolitan area. Compliance with federal standards for air quality (Table 2) is requiring the implementation of several strategies and programs to reduce emissions from both industry and vehicles.

2.3 Acoustics

Noise levels were monitored on the site between 4:30 and 6:30 P.M. on Monday, February 26, 1973. Readings were taken near the Golden Age Home since that institution is the most sensitive factor related to the project in terms of acoustical impact.

A Bruel and Kjaer sound level meter with a 1-inch microphone and windscreen were used.

During the peak travel hour a background level of 55dBA was recorded. Although the traffic on 101 was audible, it did not create any significant decibel increase. No other permanent noise sources were identified; however, approximately 2,000 feet away from the Home, a building was being razed. The effect of this operation on the background noise level was imperceptible.

---

[3] Air Quality Services Laboratory, City of Brunswick.

Table 1. Monthly Summary of Pollutant Measures
City of Brunswick
1971

	NO (1)	NO₂ (1)	SO₂ (1)	THC (1)	CO (1)	TOₓ (1)	Sus. Dust (2)	Soil Index (3)	Sulf. Index (4)	Sett. Dust (4)	Sulf. Index (5)	Sett. Dust
									OLD Network		NEW Network	
Jan.	.133	.047	.048	2.6	4.1	.001	115	1.26	2.09	62.4	1.09	28.6
Feb.	.076	.056	.032	2.3	3.3	.008	115	1.27	1.70	39.6	1.41	34.7
Mar.	.041	.032	.021	2.1	3.3	.013	94	1.05	1.10	36.4	0.99	33.3
Apr.	.035	.032	.010	2.2	3.6	.013	105	1.02	0.88	50.0	0.61	34.3
May	.025	.039	.027	2.2	3.6	.012	92	0.82	0.73	31.2	0.57	24.4
June	.020	.046	.040	2.4	2.8	.016	100	0.71	0.71	28.2	0.79	18.8
July	.021	.028	.026	2.4	2.4	.017	82	0.60	0.69	22.1	0.63	20.8
Aug.	.021	.029	.018	2.4	3.0	.015	113	0.80	0.35	25.7	0.28	23.5
Sept.	.021	.030	.016	2.3	2.8	.012	101	1.07	0.57	22.6	0.46	23.9
Oct.	.054	.037	.024	2.5	4.1	.012	123	1.14	0.70	21.2	0.68	19.0
Nov.	.043	.023	.029	2.4	2.9	.011	96	1.22	1.25	21.6	0.90	20.7
Dec.	.045	.022	.042	2.6	2.8	.006	107	1.03	1.38	29.2	1.03	26.2
YEAR	.045	.035	.028	2.4	3.2	.011	104	1.00	1.01	32.5	0.79	25.7

Sample Draft Environmental Statement

Table 2. Federal Air Quality Standards

Federal air quality standards for total oxidant, oxides of nitrogen, carbon monoxide, and hydrocarbons. (Primary and secondary standards are similar for those pollutants).

Pollutant	Federal Air Quality Standard
Total Oxidant	0.08 ppm for 1 hour not to be exceeded more than once per year.
Oxides of Nitrogen	0.05 ppm annual average
Carbon Monoxide	9.0 ppm for 8 hours 35.0 ppm for 1 hour
Hydrocarbons (non-methane)	0.24 ppm for a 3 hour period from 6 A.M. to 9 A.M.

2.4 Water Quality

Within the study area, the Columbiana River is the chief source of regional public and industrial water in New Jersey, whereas ground water is the primary source on the Pennsylvania side of the river.

The city of Brunswick's Chesville intake is just upriver from the proposed project. It supplies about 60% of the city's water supply. Waters from Foster Creek and the Columbiana, below the creek, are not used for public supply; however, their quality influences the water withdrawn at the filter plant because of the approximately two-mile upstream tidal flows of the Columbiana.

Discharge of the Foster at the Front Avenue Gaging Station has ranged from a low of 6 CFS to a high of 5,160 CFS between June, 1965 and September, 1970. The closer to the confluence with the Columbiana, the greater the effect of tidal flows on the levels and discharges of Foster Creek. These variations make interpretation of the creek's contribution to the Columbiana somewhat ambiguous.

Water quality characteristics of the Foster at Front Avenue and the Columbiana at stations 37, 14, and 13 are presented in Tables 3 and 4.

Permissible and desirable quality levels for potable water supplies are shown in Table 5.

Evaluation indicates that both are polluted. Under most conditions, however, the Chesville Filter Plant is capable of improving water quality to acceptable levels in relation to BOD or COD, turbidity and suspended solids, phosphate, ammonia, nitrogen and total and fecal coliforms.

Substances such as phenols and other trace organics and inorganics including: arsenic, barium, cadimum, chromium (6+), cyanide, lead, manganese, selenium, and silver, among others, are probably present in the Columbiana and are often not totally removed during treatment. As

Table 3. Surface Water Analysis *

Foster Creek, Water Quality at Front Avenue Gaging Station (USGS) at minimum and maximum discharge rate (11/9/70–10/4/71)

Date	11/9/70	9/31/71
Time	900	902
Dischg (cfs)	20.5	1690
T (°C)	9.5	22
DO	10.4	8.1
BOD	3.8	7.5
COD	14.3	68.5
TOC	8.0	10
Sp Cond	330	110
TDS	226	84
Susp. Solids	8	274
Cl	31	5.5
Color	70	830
Turbidity	25	600
pH	—	6.7
PO_4 (Total)	5.78	1.28
NH_3–N	.64	7.8
NO_2–N	.11	.04
NO_3–N	.85	.29
Total colif	2,400/100ml	130,000/100ml
Fecal colif	0	17,800/100ml

* U.S. Geol. Survey.

well as being a potential health hazard, trace organics can result in unpleasant tastes and odors in chlorinated drinking water.

Major surface supply pumping stations and treatment plants preclude the use of ground water for drinking in Brunswick. What little is drawn serves for cooling and other industrial uses, as there has been an increased loss of ground water quality due to sewer leaks, leaching from solid waste disposal, and industrial wells and tile fields for the underground disposal of liquid wastes.[4] In addition, buildings, extensive paving and movement of runoff from storm sewer systems prevent normal dilution and oxidation. These activities result in total iron levels from 0.08 to 429 ppm (usually greater than 1 ppm); dissolved solids from 135 to 4,270 ppm, (mean of 679/ppm) DO levels below 1 ppm and hardness values usually above 150 ppm.[5]

[4] Greenman and Others, 1961 Pa. Geol. Survey Bull, W13, 1965.
[5] Greenman and Others, 1961 N.J. *Geol. Survey Bull.*, W13, 1965.

Sample Draft Environmental Statement

Table 4. SURFACE WATER ANALYSIS: COLUMBIANA RIVER*

Parameter	Near Chesville Intake (No. 14)		Smith Forge (No. 13)	
Date	3/1/72	8/30/72	3/1/72	8/30/72
Time	10:33	11:20	10:25	11:09
Stage	3	2	3	2
T (°C)	4.0	27.0	4.5	27.0
pH	7.2	7.1	7.0	7.1
ALK($CaCO_3$)	35	46	34	46
Turbidity	8	13	8	12
DO	12.8	2.9	12.8	3.4
BCD	0.6	6.2	0.5	6.4
COD	10.4	10.2	—	—
Sp Cond.	195	225	200	225
Cl^1	18	12	18	12
PO_4 (ortho)	.35	.30	.34	.29
PO_4 (poly)	.10	.06	.09	.05
NH_3–N	.69	.12	.64	.11
NO_2–N	.029	.085	.029	.072
NO_2–N	.77	1.86	.82	1.77
Phenol	.003	.003	—	—
MBAS	.08	.04	—	—
Total colif	1,600/100ml	7,000/100ml	2,200/100ml	8,000/100ml
Fecal colif	50/100ml	130/100ml	50/100ml	390/100ml

	At Chesville Intake (No. 37) †			
Parameter	Max. daily		Min. daily	
Sp. cond. (1960-69)	570		78	
DO (1961-69)	14.5		0.0 (on many drought days)	

* City of Brunswick Water Dept.
† U.S. Geol. Survey.

The content of water pumped from the Valley Forge Co. well is shown on Table 6. This analysis is typical of such groundwater, better than most subsurface water found nearer the Center City area.

Groundwater is the major source of public supply in Pennsylvania opposite the project site. Both public and industrial supplies are located within a two mile radius of the proposed construction (Figure 2 shows several such wells). The shallow wells (50 feet or less) adjacent to the Columbiana River obtain groundwater by inducing infiltration from the river. The wells most vulnerable to this effect are deep wells located

Table 5. Criteria for Public Water Supplies *

A and VA denote absent and virtually absent respectively

Constituent or Characteristic	Permissible Criteria	Desirable Criteria
Color	75	<10
Turbidity	—	VA
Total colif	10,000/100ml	<100/100ml
Fecal colif	2,000/100ml	<20/100ml
NH_3–N	.5	<.01
Cl	250	<25
Cr^{6+}	.05	A
Cu	1.0	VA
DO	4 (monthly mean) 3 (indiv. sample)	
ALK($CaCO_3$)	—	>30-500
F	—	<0.8 to 1.7 (Temp. dependent)
Hardness ($CaCO_3$)	—	<about 150
Fo	0.3	VA
Pb	.05	A
NO_3 & NO_2–N	10	VA
PO_4	—	<about 0.1
SO_4	250	<50
TDS	500	<200
Sp Cond. (Based on TDS)	750	<300
Zn	5	VA
MBAS	.5	VA
Phenols	.001	A

   * Based on 'Water Quality Criteria' April 1, 1968, Federal Water Pollution Control Admin., Table II-1 and Text.

adjacent to the river. The migrating pollution has contributed to a loss of quality in the Center City area.[6] This movement results from the flow pattern adjacent to and beneath the Columbiana which is set up by local geology described in the section devoted to local geological conditions. Clayey silt does not readily transmit groundwater, but rather insulates underlying sediments from receiving more shallow recharge. These underlying sediments are relatively permeable sands and gravels with occasional beds of clay or silt which yield as much as 1,500 GPM to wells in Early County, Pennsylvania, and as late as 1968, it supplied most of the 26/MGD of groundwater used in that county.

Groundwater flows in response to hydrostatic gradients. Data taken

   [6] Langmuir, 1969, Penna. Water Resources (Circa 19).

Table 6. WELL WATER ANALYSIS

Well	Valley Forge Co.	Summit Water Co. No. 1	Summit Water Co. No. 2	Evca Sand & Gravel	Great Sponge Iron Co.
MSL depth (ft)	39	42	83	22	99
MSL intake internal (ft)	29-39	?-42	57-83	112-22	84-99
Date	8/21/67	4/30/64	8/18/66	10/28/65	10/28/65
Silica	20	—	5.8		
Fe (Total)	5.8	—	.06	>.5	<.05
Mn (Total)	2.2	—	.77		
Ca	44	—	15		
Mg	21	—	8.4		
Na	31	20	9.5		
K	3.1	—	3.5		
ALK	240	62	54		
SO_4	23	36	29		
Cl	36	10	12		
F	.4	—	.0		
PO_4 (Total)	.1	—	.0		
MBAS	.1	—	.0		
NO_3	.8	15	12		
Sp. cond.	527	236	209	161	68
PH	6.6	6.8	6.4		
Color	5	2	5		
Hardness ($CaCO_3$)	197	89	100		
T (°F)	62	57	56		54
Date	—	3/29/68			
Sp. Cond	—	462			
Date	—	8/24/69			
Sp. Cond.	—	384			

from wells near the proposed project (Table 7) show that static water levels exist within a few feet of mean sea level, but drop several tens of feet during pumping operations. Thus, at present, there is no widespread migration of groundwater to Pennsylvania, although some movement may occur during heavy pumping periods in Pennsylvania. This type of activity could increase in the future when greater demand results in heavier pumping activity.

Dredging of the Columbiana probably short circuits this type of migration and induces groundwater recharge by river water into deep aquifer until the dredged areas are again layered with river silt. However, at this time, there is no local evidence that river water quality has adversely affected groundwater quality on either side of the river.

A-16   Sample Draft Environmental Statement

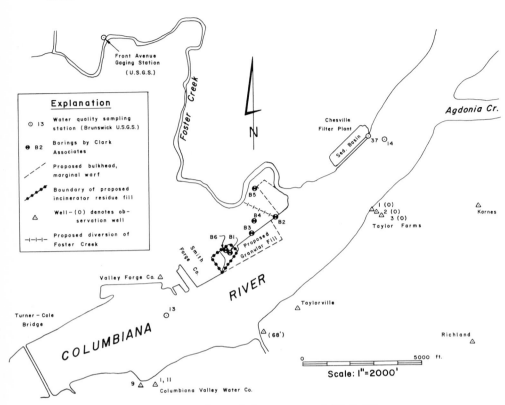

Figure 2 General Location and Well Sites.

Part of the area under consideration, located southwest of the Foster Creek mouth, is being used as a landfill site. Between 1960 and 1970 the field was used as an open dump. In 1971 the city excavated pits for disposal of residues from the city incinerators.[7] Residues are disposed into water ponded in the pits, which is continuous with adjacent groundwater. The specific conductance of water adjacent to the residue at one such pit was 2,090 mhos on January 28, 1973. This value would have been much higher had the water not been diluted by both rainwater and surface runoff. Table 8 gives an analysis of incinerator residue leachate typical of the leachate formed under these conditions. At least a small amount of this material must seep into the Columbiana River and Foster Creek.

[7] Industrial wastes are not processed by city installations.

Table 7. GROUND WATER LEVELS

Well or Boring No.	B1	B2	B3	B4	B5	B6	Valley Forge Co.	Summit Water Co. No. 9	Summit Water Co. No. 11	Cinnaminson 68' Well
Date							10/68	10/68	10/68	10/68
MSL depth (ft.)	−39	−23	−30				−39		−88	−62 (bedrock)
Intake internal below MSL							29-39	39-60	57-83	19-62
MSL static level	−2	−2	−5	−2	−6	−2	−1	−5	−6	
MSL pumping level								−27		
Pump rate (gpm)							310	503		350

Well	Maurer Farms	EVCA	Great Sponge
Date	10/68	10/68	10/68
MSL depth (ft.)	−10	−26	−99
Intake internal below MSL	5-10	10-26	64-99
MSL pumping level			−49
MSL static level	+6	+4	+19
Pump rate or yield			326

Table 8. Quality of Leachate from Incinerator Residue of Municipal Solid Wastes

PH	8.3
ALK ($CaCO_3$)	4,260
NO_3–N	3.5
PO_4	15.2
Fe	5.0
Cl	1,800
F	5
SO_4	94
Ca	21
Na	3,350
K	22
Total dissolved solids	7,933
BOD	125
COD	1,265
Zn	1.0
Cr (Total)	1.5
Cu	1.2
Pb	1.2
NH_3	47.6
Kjeldahl Nitrogen	125
Sp. Cond. (Estimated)	12,200

(After Neiswiender and Mellinger, *J. Soil Mechanics and Foundations Div.*, Proc. Am. Soc. Civ. Eng., Oct., 1971, p. 8451.)

2.5 Geology

A typical geological cross-section (Figure 3), based on data and well log information provided by the U. S. Geological Survey indicates that the land surface is mantled by a layer of clayey silt and fill, with some sand of flood plain origin probably of recent or Pleistocene Age. This layer is missing or else has been removed in certain locations. Similar material is found lining the bed of the Columbiana River except where it has been removed by dredging activity.

Underlying sediments consist of relatively permeable sands and gravels with occasional beds of clay or silt. These may be the Cape May Formation, or Pensauken Formation, both of Quaternary Age, or the Magothy and Raritan Formation, which is Cretaceous in age. These in turn rest on the impermeable Wissahickon Schist.

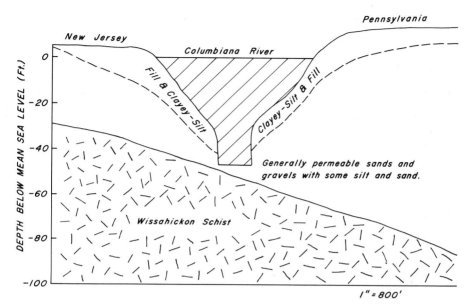

Figure 3 Cross Sections of Localized Typical Geology.

2.6 Terrestrial Ecology

Municipal ownership and limited construction have resulted in several open, natural, or seminatural pockets within the area associated by this project. They include:

a. The very limited park area of the Golden Age Home, which includes a row of old maple trees aligned at right angles to the river. The ground cover consists of a combination of weeds and grass.

b. The area that was farmed by the Vo-Tech School until recently. Although not currently planted, crop marks are still visible on aerial photographs. Some of this area is being used as the landfill site.

c. Lower Foster Creek from State Road to the Columbiana River. Backfilling to the edge of the Creek has eliminated most of the tidal marsh; however, the Creek edges are still relatively undisturbed. Emergent aquatic vegetation includes in various degrees of number:
Cat Tail (Typha litifolia);
Tree of Heaven (Ailanthus altissma);
Princess Tree (Paulownia tomentosa).

d. The dense willow forest of 12 acres found on the site of an old

filtration bed approximately 20 feet above the river elevation. Major species include:
 Willow (Salix sp.);
 Poplar (Populus sp.);
 Red Mulberry (Morus rubra).
Since such small wooded areas are almost nonexistent for many miles on the New Jersey side of the Columbiana, it is used as a habitat for these groups of birds:
 Pheasants;
 A broad variety of migratory species following the Columbiana;
 Winter residents including the white throated sparrow (Zonotrichia albicollis);
 Overwintering species that usually migrate south including: the red-eyed towee (Pipilo crythrophthalmus) and the mourning dove (Zenaldura macroura).

 e. The natural riverfront, which is separated by a dyke from the wooded area described above, which contained the abandoned filtration bed. This section of the river is lined with mature (older than 50 years) black willow (Salix nigra).

 f. A small area surrounding the abandoned pumphouse containing approximately 24 mature (older than 80 years) sycamore trees (Plantanus accidentalis) that form an open grove on the river's edge.

Although no rare or endangered species exist, these last three comprise one of the few balanced sites in this area where one can actually stand in a natural setting.

2.7 Aquatic Biology

The aquatic life of the Columbiana River in the vicinity of Chesville is listed in Table 9, and the species in Foster Creek in Table 10. These data were summarized from studies conducted from 1957 to 1971.[8,9]

The diversity of plant and animal life is not indicative of clean stream ecosystems in either the Columbiana or the Foster. In a healthy stream system macroinvertebrates are represented by as many as 49 taxa, while in the Columbiana only 2 taxa have been reported, and these were classified as being pollution tolerant. Foster is reported to have 7

---

[8] Ecological Study of Columbiana River in the vicinity of North Island, Ichthyological Assoc., Borgville, N.J. 1971.

[9] Dept. of Limnology, 1959. Biological Studies of Columbiana River for the Interstate Commission on Columbiana River 1957–59. Academy of Sciences of Brunswick, N.J.

Table 9. LIST OF BIOTA REPORTED AS PRESENT IN THE COLUMBIANA IN THE VICINITY OF THE PROPOSED PIER CONSTRUCTION

Scientific Name	Common Name
Aufwuchs (only most abundant species are listed)—National Water Quality—1963	
	Diatoms
Nitzschia acicularis	
Nitzschia tryblionella	
Nitzschia denticula	
Synedra ulna	
Dysloneis smithii	
Rhizosolenia eriensis	
Surirella sp.	
Plankton	Rotifors
Keratella	
Brachionus	
Trichocerca	
Benthic Macroinvertebrates (Zemaitis and Cos, 1971)	
	Tubificid worms
Limnodrilus ceroix	
Limnodrilus hoffmeisteri	
Fishes (Anselmini, 1971)	
Alosa aestivalis	Blueback herring
Alosa pseudoharengus	American shad
Cyprinus carpio	Carp
Hybognathus nuchalis	Gudgeon
Notemigonus crysoleucas	Golden shiner
Notropis analostanus	Satinfin shiner
Notropis hudsonius	Spottail shiner
Ictalurus catus	White catfish
Ictalurus nebulosus	Brown bullhead
Ictalurus punctatus	Channel catfish
Anguilla rostrata	Common eel
Fundulus diaphanus	Freshwater killifish
Fundulus heteroclitus	Common killifish
Morone americana	White perch
Lepomis auritus	Yellowbelly sunfish
Etheostoma olmstedi	Tessellated darter

macroinvertebrate taxa, most of which are also tolerant species. Therefore both streams may be currently classified as severely stressed in relation to their ecosystems.

The fish population consists of either pollution-tolerant herring, which are only found in the lower Columbiana River and Foster Creek briefly during the migratory seasons.

Table 10. Species List Reported from Foster Creek, Brunswick, N.J.

Scientific Name	Common Name
Plants	
	Blue-green algae
Gomphosphaeria	
Oscillitoria	
	Desmids
Scenedesmus	
Closterium	
Cosmarium	
Ulothrix	
Stigeoclonium	
	Diatoms
Diatoma	
Rhopalodia	
Navicula	
Gomphonema	
Melosira	
Elodea canadensis	Elodea
Fissidens	Moss
Animals	
Oligochaetae	
Tubifex	Tubificid worm
Lumbricius	Aquatic worm
Arthropoda	
Ephemerella sp.	Mayfly
Peltoperia sp.	Stonefly
Isopoda	Sow bug
Hydrophysche sp.	Caddis fly
Tendipedidae sp. (2)	Midge
Pisces	
Cyprinus carpio	Carp
Notemigonus crysoleucas	Golden shiner
Notropis hudsonius	Spottail shiner
Notropis amoenus	Comely shiner
Rhinichythyes cataractae	Long-nosed dace
Ictalurus nebulosus	Brown bullhead
Anguilla rostrata	American eel
Fundulus diaphanus	Freshwater killifish
Lepomis auritus	Yellowbelly sunfish

Among the equatic flora and fauna, there are no rare or endangered species represented.

2.8 Socio-Economics

The area affected by the proposed project is city owned, abutted only by the Vo-Tech School, House of Detention, and the Golden Age Home. With the exception of a limited wooded area, the surroundings, although open, are not aesthetically sensitive in character.

At present the area has no commercial application and produces no revenue. It is owned by the city and therefore is not a source of income from taxation. The nature of the property and the surrounding land use is such that it is doubtful that there will be any significant change in these conditions within the foreseeable future.

The three municipal institutions establish the parameters of the local population and a framework for ensuing social effects. One is a vocational school, and another is of a disciplinary nature used for limited periods of internment. The third, however, is a home for the elderly which should be reviewed for potential impact on the residents, who utilize the facility as their permanent and last residence.

2.9 Historic/Archeological Sites

Although no historic sites are found in the project area, one listed on both the National and State Register is located approximately 3,000 feet northwest of the proposed project. The Lambert House was built in 1753. It is in excellent condition due to restoration by the State Museum. Presently it serves as a local museum and houses the offices of the Brunswick Historical Society.

In addition to the Federal Register and the state listing, county and local historical groups were contacted to determine if any other similar resources were located near the project. None were to be found.

There are no known archeological dig sites in the project boundaries, which was verified by a conversation with the State Archeologist.

3. PROPOSED ACTION AND ALTERNATIVES

3.1 Project Description

The city of Brunswick, through the Brunswick Port Authority, is planning to expand its marine facilities by developing a new terminal facility on the Brunswick side of the Columbiana River.

The project will be known as the Jennert Marine Terminal and will utilize a tract of city-owned land adjacent to Hough Street, the Golden Age Home and the Vo-Tech School, all within the city limits.

Initially it will be used as an unloading and storage site for automobile imports. It is ultimately intended to be an important and versatile segment of an extensive system of marine facilities developed by the city. When completed, the terminal will encompass approximately 160 acres. However, it is proposed as a phased project with three separate stages of construction.

Phase 1 (Figure 4) The initial stage will involve the construction of a rock embankment starting 70 feet inshore of the pierhead line at the upstream (northeast) property line of Smith Forge Company and extending upstream (northeast) 1,100 feet to a point where it intersects the tip of a small peninsula located inshore of the bulkhead line. Two vacant slips are included. After the embankment is completed, the two slip areas (7.0 acres) and the adjacent low-lying land (6.2 acres) will be filled even with the surrounding ground level. With existing available land, this phase will yield approximatey 80 usable acres when completed.

Phase 2 (Figures 4 and 5) The second stage will include construction of marginal berthing facilities that will be part of a 95-acre marine terminal. Forty of these acres will be created by filling shoreward from a proposed bulkhead that will form the berthing. The bulkhead will be 2,430 feet in length and extend upstream from the existing Northern Metal Company wharf on the same alignment. This will result in an encroachment of 395 feet into the river from the Federal Pierhead Line at the extreme upstream corner of the wharf.

Phase 3 (Figure 4) The third stage will involve another bulkhead beginning at the western edge of the Foster Creek mouth, closing the mouth of the Creek and extending upstream 2,300 feet to a point just southwest of Foster Street and the Chesville Filter Plant settling basin. The purpose of this phase is to generate maximum usage for the area disected by the meanders of the creek. It is then proposed that the area behind the bulkhead be filled. This includes the mouth and approximately 1,800 feet of the Foster Creek channel. A new outlet will be constructed directly upstream of the marginal wharf proposed under Phase 2. This new covered channel, 1,400 feet in length, will be 133 feet wide at the bottom and 178 feet wide at the underside of the cover.

The section, when completed, will be converted to become productive, which is significant because of the limited amount of available river front property and its value.

3.2 Engineering Data (Figures 6 & 7)

In Phase 1, approximately 80,000 cubic yards of 14 feet thick silt will have to be excavated so that the proposed dyke will be based on a hard granular strata, which is found at minus 17.5 elevation. The material

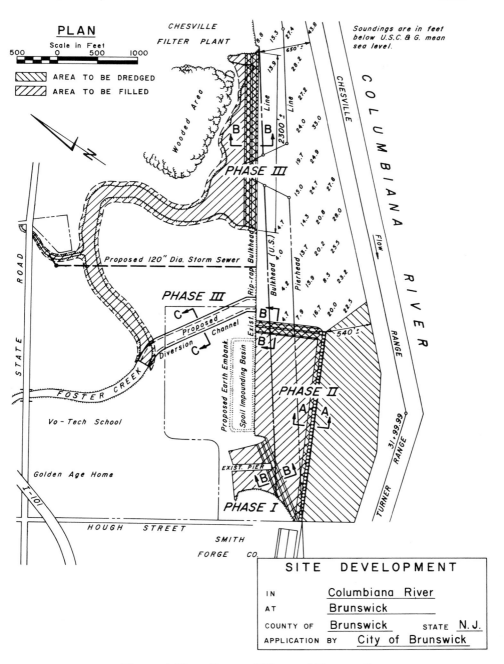

Figure 4 Three Proposed Phases of Construction.

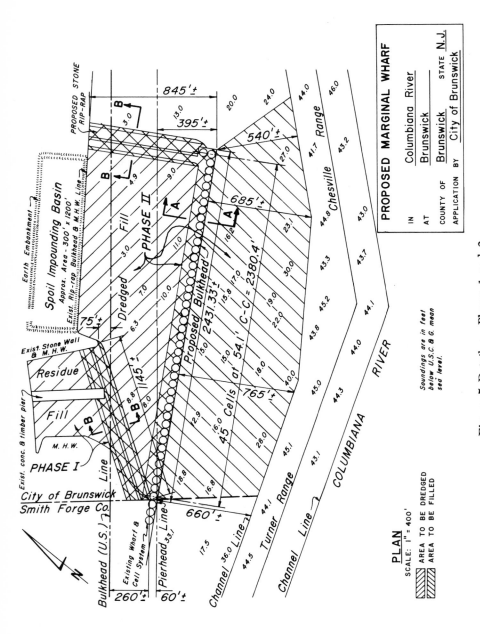

Figure 5 Detail on Phases 1 and 2.

Sample Draft Environmental Statement

will be removed by dragline or bucket-equipped cranes and deposited inshore on ground contained within a continuous earth movement adjacent to the fill area. The rock for the embankment will be back-dumped from trucks that will use the same areas as the cranes involved in silt removal, advancing out on the embankment as it is created. About 72,600 cubic

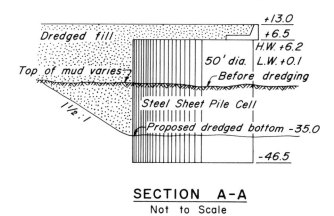

SECTION A-A
Not to Scale
PHASE II

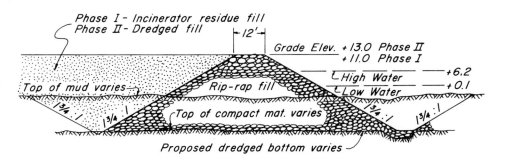

SECTION B-B
Not to Scale
Stone Rip-rap Construction Similar For Phases I, II, & III

PROPOSED MARGINAL WHARF	
IN	Columbiana River
AT	Brunswick
COUNTY OF	Brunswick STATE N. J.
APPLICATION BY	City of Brunswick

Figure 6 Construction Sections (A-A, B-B).

yards of rock-fill will be required for the embankment. If this material is not readily available, brickbats and concrete rubble obtained from demolitions will be substituted.

The silt and clay previously removed, topped by select incinerator refuse from city installations, will be used to elevate the area to the same

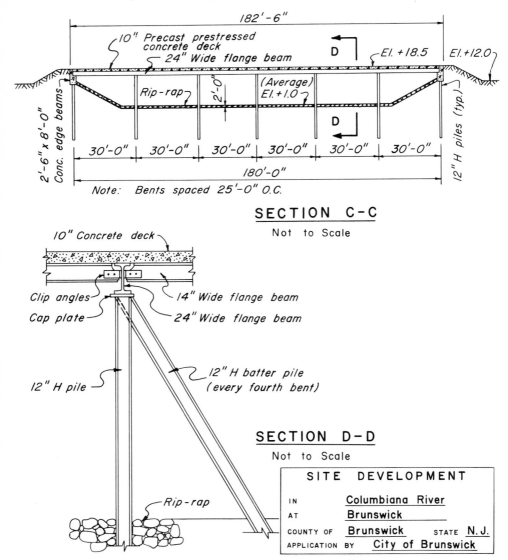

Figure 7 Construction Sections (C-C, D-D).

Sample Draft Environmental Statement

A-29

grade level as the embankment—11.0 feet.[10] Approximately 120,000 cubic yards of fill will be required, of which 2/3 will be dredged material and 1/3 incinerator refuse.

In Phase 2 the proposed bulkhead will be constructed of sheet pile cells 50.93 feet in diameter fronted by a reinforced concrete sea wall. The cell construction will require predredging to an elevation of minus 30 feet. A rip-rapped embankment will be constructed on the perpendicular for the entire length to an existing embankment of similar construction facing the river boundary.

Dredging to minus 35 feet will be completed channel-side of the embankment out to the existing channel. This dredging will result in the removal of approximately 675,000 cubic yards of unsuitable silt and clay.

An estimated 1,500,000 cubic yards of fill will be required to complete the project. Suitable and approved material will be obained by the contractor from his sources and those suggested by the Corps of Engineers. All necessary permits will be obtained and there will be compliance with all regulations.

In Phase 3 the invert elevation of the reconstructed channel will coincide with the elevation of the creek bed at the point of diversion and will be sloped to carry peak flows.

The channel will be covered with prestressed concrete deck slabs supported by structural steel framing. That system in turn will be based on steel H piles. Reinforced concrete headwall and outfall structures will be constructed to prevent erosion. The bottom and sides of the channel will be protected by 2 feet of graded rip-rap.

Cross sectional data indicates a maximum flow of 23,800 CFS under the State Road bridge before damming action begins. The new channel will be designed to carry an additional 25 per cent capacity.

Approximately 42,000 cubic yards of silt will have to be dredged from the outfall out to where the river is sufficiently deep so that the flow from the outfall will not be impeded. Again, this will be done by bucket and disposed at an approved site.

110,000 cubic yards of silt, later used as fill, will have to be excavated for the construction of the channel. The former creek bed will be filled to a grade of 16.0, Corps of Engineer Datum. This will require approximately 720,000 cubic yards of material, of which a considerable percentage will be incinerator residue.

The concept of enclosing the existing channel in a culvert and paving the surface was examined and rejected as impractical from the standpoints of both load requirements and economics. The channel area, in order to have a comparable load capacity, will require vertical support, which makes the approach unfeasible. In addition, the proposed realign-

[10] United States Engineering Datum; Brunswick to Monthouth.

ment is 1,400 feet as compared with the 4,600 feet existing alignment, over three times the length.

Around 160,000 cubic yards of silt will have to be removed for construction of the embankment. The disposal site will have been approved.

By phase, estimated construction costs will amount to:

Phase 1 — $2,283,000.00;
Phase 2 — $7,519,000,00;
Phase 3 — $6,077,000.00.

3.3 Alternatives

Because the project will require use of specific city-owned property previously described, there are no alternative sites that can be considered. Therefore, the only other options are:

1. Eliminate Phase 2;
2. Eliminate Phase 3;
3. Expand existing facilities to meet demand;
4. No Action.

The construction of Phase 1 would result in a project totaling approximately 80 acres. If Phase 2 were completed the project would total 95 acres. Phase 3 represents an addition of approximately 65 acres or a total of 160 acres for the entire complex.

4. ENVIRONMENTAL IMPACTS

4.1 Climate

There will be no significant effect on climatic or meteorological factors resulting from the construction of the proposed project or any portion of it.

4.2 Air Quality

There will be no significant permanent impact on ambient air quality as a result of this project.

Temporarily, any construction operation might cause dust, as well as smoke, if burning is required. However, this will be controlled by arrangements requiring the contractor to take precautionary measures.

When construction is complete, the major source of emissions will stem from trucks moving material to and from the terminal. Initially the

Sample Draft Environmental Statement A-31

import of 100,000 cars per year is projected. Haul-a-ways will be used, and with a load of roughly 6 cars per unit it will require an average of 64 trucks per day, per five-day week. As the site is in Brunswick County, the direct effect is calculated on 12 miles to the pier from county limits and 12 miles return, or a total of 24 miles. In addition it is assumed that the vehicles will be driven from the point of unloading to a storage location. This should average approximately 300 feet (0.05 miles) per vehicle.

Polluting emissions resulting from these activities are shown in Table 11.

This does not represent a significant input. Even without the full anticipated reduction of emissions required by 1975 by federal regulations, any negative impact in relation to Air Quality will be extremely limited.

Alternatives 1 and/or 2 will represent even less emission as there will be less traffic volume. There will be some temporary increase of dust and smoke as a result of construction activities, but these can be limited by contractual arrangements and abatement measures.

Table 11. TRUCK EMISSIONS [*]

Containment	Pounds/1000 gallons	Pounds/day	Pounds/year
Solids	110	56	1,460
SO_2	45	13	600
NO_2	220	11	2,926
Hydrocarbons	320	16	4,250
Organic Acids	30	2	400
Aldehydes	16	0.8	213
Ammonia	2	0.1	27

[*] Based on diesel fuel at 4.8 gallons per round trip in Brunwick County.

AUTOMOTIVE EMISSIONS

Containment	Pounds/Day		Pounds/Year	
	1970 [a]	1976 [b]	1970 [a]	1976 [b]
Carbon Monoxide	.51	.050	132	13
NO_x	.46	.042	112	11
Total Hydrocarbons	.05	.008	15	1

[a] No emission control.
[b] With federal controls.

4.3 Acoustics

With a volume of 100,000 vehicles per year through the Jennert Terminal, the movement of these imports, either by rail, individually, or by car carriers, will result in increased decibel levels for the immediate area.

The National Bureau of standards has measured car carriers at levels varying from 80 to 90 dB(A) from the time of start up through acceleration to time of passing a monitoring microphone located 50 feet from the roadway.

Assuming less than 10 such vehicles per hour, the acoustic point source model can be used. Noise levels reduce 6 dB per doubling of distance. Therefore, at 200 feet from Hartel Road, well within the Golden Age Home and the Vo Tech School area, noise levels would vary from 68 to 78 dB(A). This would result in an increase of from 13 to 23 dB over existing conditions and would cause significant annoyance, especially during daytime rest hours. It is possible that the levels could exceed both FHWA and HUD specifications. With a greater truck density further analysis would be required as prediction schemes are based on an auto-truck mix.

With regard to unloading noise, the distance from the dock area should preclude any effect on the residents of the institutions. All increases would be proportional to the final level of construction. If existing facilities were expanded, the levels would be increased at each site.

On a temporary basis, construction operations will cause some additional noise as a result of both construction equipment and trucks transporting in both directions. This annoyance will cease with completion of the project, and constructural arrangements relating to the use of noise-quieted equipment will keep this annoyance to a minimum.

4.4 Water Quality

Most of the negative project effects on the Columbiana River are not of a permanent nature. Dredging of a new channel to a depth of 35 feet will temporarily increase suspended sediment levels as will the filling operation involving granular material. Dredging will also expose more of the permeable acquifer.

Control over dredging operations to insure against excessive suspended material will be exercised through similar technical provisions as are included in the Corps of Engineers' contractual arrangements for dredging activities. These include such safeguards as density sampling and the cessation of pumping into the disposal area whenever the density of suspended materials exceeds 8 grams per liter in excess of the river

Sample Draft Environmental Statement A-33

water taken near the dredging site. In addition, the dredging contractor will be required to observe Corps of Engineers' specifications for the marking and lighting of dredging plants.[11]

Effects of the dredging activity, however, will abate when settling occurs and there will be minor permanent degradation of the river water quality or impact on its population.

Care will be taken during the original dredging period and also during subsequent maintenance operations to take into account tidal and wind effects and to insure that the volume of sediment does not exceed the removal capacity of the Chesville Sedimentation Basin.

The removal of sludge from the river bottom could also remove deposits of contaminating trace metals, and since this unsuitable material will be dumped, after approval by responsible agencies, at a selected site away from the river, water quality could improve in relation to those toxic materials.

The completed project will not be such that its discharges will add to the river pollution levels.

The use of incinerator refuse could create a negative impact. However, this material will be used only above the water saturated zone. Proper compaction will be assured by contractual arrangements, continuous inspection, and paving with an impervious material to prevent leeching action.

The quality of the refuse will be much improved over current material as the older, less efficient incinerator installations are shortly scheduled to be shut down by the city.

Finally, a drain system will be installed to assure that runoff will be efficiently carried away from the site. If necessary, sealing material can be used, although a recent article [12] suggests that not only can leachate problems be controlled, but they do not have the negative effects heretofore believed. Utilizing the procedures outlined above and the cessation of the uncontrolled activities should actually result in localized improvement in water quality.

To further combat reduction in water quality, sewage reception pits to receive stored shipboard wastes will be provided at the wharf face. These will be tied into the terminal sanitary sewer system, the city system, and will ultimately be treated at city sanitation plants.

Rechannelization of the mouth of Foster Creek will have a localized effect on natural stream action and the aquatic ecosystem.

Channelization almost always creates some negative effects. The elimination of the meanders removes the extra floodplain area needed

[11] C.F.R. Title 33, Part 201.
[12] *Civil Engineering—ASCE*, March 1973, pp. 69–71.

during flood conditions. Thus, the new channel will be designed to insure against the possibility of water backing upstream under flooding conditions; in fact volume capacity will be equal to or better than that of the bridge site further upstream.

This project will also include some access for periodic removal of accumulated sediment and will allow for high flows to avoid flooding.

Consideration is being given to devoting approximately 8 to 12 acres of riverside woodland for a municipal waterfront park.

Restricted construction would produce the same type of effects but to a lesser degree such as reducing the amount of dredging activity. However, not completing Section 3 would preclude channelization of the Foster Creek, eliminating the projected impacts on water quality and aquatic life. Improving existing sites would have the same effects at several locations.

During the construction period there will be greater turbidity because of various construction activities beyond dredging. However, these will also be of a temporary nature and this condition will be alleviated after completion.

4.5 Geology

Dredging activity could expose more of the permeable deep aquifer to river discharge; however, if this condition is monitored, and river water quality does not decline further, the impact will not be significant.

No other impacts are projected.

4.6 Terrestrial Ecology

The first phase of this project will have no significant impact on factors relating to terrestrial ecology. The lack of flora or fauna precludes any loss, with the exception of a small grass area containing a very few trees.

The second phase will also have relatively little significant impact on ecological factors. The open fields formerly used for farming that now lie fallow would be the only loss.

Phase three will have a much greater negative impact. The destruction of the wooded area located on the old filtration beds site will result in the loss of one of the last vestiges of natural riverbank for an extended distance downstream. Although the wildlife is limited, it will be required to seek new habitats, upstream from the mouth area of the creek.

Improving existing facilities would have little effect on this factor; as would no-action.

4.7 Aquatic Biology

Impacts of the project on aquatic biota will be primarily related to:

1. Direct effects of dredging on bottom life;
2. Effects of increased turbidity on fish and benthic forms;
3. Effects of rechannelization of Foster Creek on indigenous biota;
4. The effect of dredging and rechannelization on migratory fishes.

Dredging should have little or no permanent effect on the existing fish population. It will cause them to move out of the area during the period of dredging operations, however this will be temporary. Bottom fauna (two species of tubificid annelids) will not be affected because of their extreme resilience. The temporary loss of benthic organisms during the construction period will not permanently alter the food web relationships.

Excessive turbidity could cause problems by affecting the respiratory processes of some aquatic organisms, therefore, efforts will be made to keep turbidity levels to a minimum. A recent article by Dr. Joel F. Gustafson [13] suggests that dredging turbidity plays a positive role in removing toxic material from the water column by flocculation. In still another article [14] Dr. Gustafson suggests that many of the negative impacts associated with dredging are exaggerated and that the entire matter merits more study to establish ultimate effects.

The rechannelization of lower Foster Creek could result in a change in the aquatic biota in that area. Often the artificial channel establishes a homogeneous substrate which impedes species diversification. However, the water quality is such that the number and variety of species is and will be limited.

There is insufficient data at this time to substantiate that shad and herring spawn in the Foster. It is indicated, however, that in the future there will be an improvement in the water quality of the creek. In that event, it is believed that river herring and bass will migrate upstream to spawn. The new channel will not affect the migrations, but will preclude what would normally be a nursery area at the mouth of the creek. If Phase 3 is not constructed there will be minimal impact on Foster Creek. Improvement on existing facilities would have localized effects as described above. No-action would have no negative impact.

It has been determined that significant numbers of shad (Alosa

---

[13] Joel F. Gustafson, "Beneficial Effects of Dredging Turbidity," *World Dredging and Marine Construction*, Dec. 1972.

[14] Joel F. Gustafson, "Ecological Effects of Dredged Barrow Pits," *World Dredging and Marine Construction*, Sept. 1972.

pseudeharengus) and herring (Alosa aestivalis) are present in the lower Columbiana River during the spring upstream run from about March 15 to June 15 and the fall downstream run from about September 15 to November 15. There have been no specific studies addressed to the net effects of wharf construction on these species; however, the results at this time are considered detrimental; to what degree is unknown.

4.8 Socio-Economics

The direct negative sociological impacts of the Jennert project will relate primarily to the Golden Age Home. The residents will lose the small area on the waterfront reserved for their use. They will also be temporarily subjected to those nuisances created by major construction, including increased noise levels, dust, heavier traffic on Hazel Street, and other similar effects. On a permanent basis, the traffic resulting from terminal activities will cause some disturbances, which will be directly proportional to the type and volume of material handled by the installation.

Those Vo Tech School buildings directly adjoining Hazel Street will also be subjected to the effects of traffic on that thoroughfare.

There will be no significant displacement of or change in neighborhood composition.

From the negative economic standpoint, there will be some direct costs related to additional public services such as water, sewer, and traffic control, among others, required by the installation. Jennert Terminal, however, also offers several beneficial impacts on both a temporary and permanent basis.

Construction will require labor of varying trades and skill levels. Approximately 15 million dollars will be placed into the economy of which the largest share will be on a local basis. Beyond that, at least a 1.2 million dollar maritime payroll (200± employees) would result from the projected import volume of 100,000 vehicles per year. A large percentage of the labor would require no specific skill or training, which would provide work for those in our society who are finding it most difficult to obtain it. This could have some impact on welfare and unemployment compensation.

Estimates project that the facility will spend about 4.85 million dollars per year on direct purchases. This money should also be reflected in the local economy.

The result of the money put into circulation by the implementation of the project will be continued circulation among the recipients. This is called the multiplier effect. Without an extensive study, a factor of 2(x) would be a conservative figure. Based on the 4.85 million dollars

Sample Draft Environmental Statement

spent each year, the total direct and indirect contributions of the facility to the Brunswick economy will total 9.7 million dollars per annum.

The facility, whether it is operated by the Brunswick Port Authority or leased as an operating concession, will generate revenue for the city (Table 12). These funds will not only compensate for construction costs, but will act as a long-term source of revenue.

The facility, by virtue of its total marginal berth of 3,600 feet, will allow for the docking of the largest freighters in operation (±900 feet). Beyond the more obvious benefits lies another basic advantage to this project. Several metropolitan ports are in strong competition to attract maritime clientele to their respective areas. It is a known marketing principal that firmly established, better equipped operations in a competitive situation are more apt to generate additional trade than underequipped, marginal operations. This concept would certainly apply in this situation. All other conditions being equal, the construction of this project would tend to attract new business for the Brunswick Port facilities by virtue of the tonnage capacity, equipment, and resulting trade relationships.

If the Jennert Terminal were not constructed there would be no effects, either positive or negative, on physical environmental factors. In addition, no adverse impact would be suffered by the adjoining city institutions. There would be, however, other significant sociological and economic losses to both the city and the general area.

The loss of this facility woud inhibit port development, the eco-

Table 12. Projected Yearly Income from Jennert Terminal

	100,000 Vehicle Level	200,000 Vehicle Level
Direct Maritime Labor Income	$ 1,240,000	$ 2,479,000
Other Direct Income,		
Including Other Labor Income	$ 2,951,968	3,546,146
Auxiliary Income	644,194	1,288,387
Sub Total:	$ 4,836,162	$ 7,313,533
State Tax Revenue	66,694	133,388
City Tax Revenue	132,963	195,191
Sub Total:	$ 199,657	328,579
Cumulative Total:	5,035,819	7,642,112
Dollar Multiplier Factor Adjustment		7,642,112
Total Annual Region Income:	$10,171,638	$15,284,224

nomic and ensuing sociological gains, and would conflict directly with established and approved local plans and goals. The lack of additional tonnage capability and revenues projected for this project will restrict the growth of port activities in the face of aggressive and expanding development by other areas in a highly competitive field.

Not proceeding with Phase 2 would considerably reduce encroachment in the Columbiana and eliminate most of the anticipated dredging activity. However, the resulting facility would fall far short of projected requirements and almost totally upset cost/revenue ratios. This will result in additional construction at a later date, when according to all forecasts, necessary expenditures of public funds will be considerably greater, reducing the feasibility of the project.

The improvement of existing facilities is not desirable in terms of economics as space is not available and the cost of acquisition on construction would be too high to make the project feasible.

4.9 Historic/Archeological Sites

Because the only historical site in the vicinity of the project is over a half mile away, and because an interstate highway separates that site from the proposed project, there will be no impact on the Lambert House.

Further, because no archeological dig sites are known to exist on the property involved, there will be no effect by implementation.

If artifacts are discovered during construction appropriate care will be given to allow for removal and the objects will be donated to the proper authorities.

4.10 Summary of Significant Effects

Jennert Terminal will develop to public advantage a section of relatively low value, city-owned property. The major benefits will be socio-economic in nature.

The primary long range effects on the physical environment include the potential loss of a wooded section near the bank of the Columbiana, and the loss of a natural channel and its associated aquatic ecological system. In spite of the proposed dredging, water quality in both the Columbiana River and Foster Creek is not particularly high, and will not be permanently affected unless caution is not exercised in the case of incinerator refuse as fill material.

Finally, residents of the Golden Age Home will be subjected to both temporary and permanent increases in noise levels which will have to be alleviated, in addition to the loss of a small recreation area. These noise levels will also affect buildings belonging to the Vo Tech School located near Hazel Avenue.

5. ADVERSE ENVIRONMENTAL EFFECTS WHICH CANNOT BE AVOIDED SHOULD PROJECT BE IMPLEMENTED

With Construction of Project:

1. Temporary
   a. Increased turbidity and siltation in the Columbiana River during both construction and maintenance dredging periods;
   b. Some minor disturbance to the deep aquifer;
   c. Temporary displacement of aquatic biota;
   d. Increase of noise levels during construction period;
   e. Some increase of dust and/or smoke during construction period.
2. Permanent
   a. An increase in acoustic levels at old folks home and vo-tech school. The increase will depend on type of attention.

With No Construction:

1. Permanent
   a. A loss of employment, revenue, and ensuing sociological and economic gains for the area;
   b. The loss of a competitive position in the port services market;
   c. No opportunity to effectively clean up the existing landfill area and the resulting leachate effects on water quality.

6. MEASURES UNDER CONSIDERATION TO MINIMIZE UNAVOIDABLE ENVIRONMENTAL EFFECTS

1. Contractual arrangements to insure against undue annoyance from excessive noise during construction by use of all possible noise-quieted equipment.
2. Contractual arrangements that include state and federal restrictions to insure against excessive air pollution during construction.
3. The erection of a suitable noise barrier to protect the residents of the Golden Age Home and the occupants of the Vo Tech School from undue annoyance.
4. Adequate landscaping to screen the noise barriers, assist in noise level reduction, and generally improve the aesthetics of the Home.
5. Investigation into the possibilities of reserving the 10–12 acres of wooded area for a public waterfront park.

6. Investigation into the development of a limited garden and/or recreational area for Golden Age or some other "trade off" for the small section of land that will be lost to the institution.

7. Strict control over the use of incinerator refuse as fill; this will encompass the depth to which it is placed, controlled compaction to assure proper density, an adequate paved cover to insure against leaching action, and a drainage system capable of controlling and directing run-off away from the site.

8. Contractual arrangements to insure against excessive turbidity and siltation during dredging operation. These efforts are to be coordinated with officials of the Chesville Filter Plant so that those facilities will be able to properly remove excessive material from the intake water without overtaxing plant capabilities.

9. Dredging operations will be scheduled so that shad and herring runs will not be disturbed in their migration.

10. Excavated unsuitable material will be disposed in a manner approved by responsible agencies.

11. Design of the new channel will insure against upstream flooding and allow access for clean-out when necessary.

12. Caution will be exercised to insure against excessive dredging of the aquifier strata so as not to affect groundwater quality.

7. THE RELATIONSHIP BETWEEN LOCAL SHORT TERM USES OF MAN'S ENVIRONMENT AND MAINTENANCE AND ENHANCEMENT OF LONG TERM PRODUCTIVITY

The construction of the proposed project will assure optimum use of a section of unused land owned by the city of Brunswick. Although it will result in a few limited temporary environmental effects, the life expectancy of the terminal's productivity will be measured in several decades and will prove profitable to society on both a regional and local basis.

8. ANY IRREVERSIBLE AND IRRETRIEVABLE COMMITMENT OF RESOURCES WHICH WOULD BE INVOLVED IN THE PROPOSED ACTION, SHOULD IT BE IMPLEMENTED

Irretrievable commitments of this project are limited to the manpower expended on design and construction, the building materials that will be used, such as concrete and steel, the power that will be used during construction, and the fuels consumed both during and after construction.

INDEX

Acoustical impacts, methods for attenuating, 51–52
Acoustics, 48, 49–52
Administrative Procedure Act, 15
Aerial photography:
  example of, 88 (fig.)
  use of, 86
Aesthetic qualities, assessment problems, 72–73
Aesthetics, as significant classification, 178
Agency coordination, importance of, 33–34
Agricultural acreage:
  destruction of, 60
  importance of, 67
Air Control Units, as data source, 39
Air pollutants, lists of, 53, 54
Air quality 48, 52–56
  monitoring programs, 52–53
Air Quality Services Laboratory, A-9fn
Alaskan Pipeline, 5
American Society of Photogrammetry, 89
Anderson, Fred R., 30
Antiquity Act (1906), 83
Aquifer, importance of safeguarding, 59–60
Archeological resources, 65
  importance of, 82–84
Assessment process:
  basic data required, 37
  data collection, 38–39
  determining levels of significance, 38
  federal guidelines study, 41
  mathematical evaluation, 39–40
  on-going surveillance of estimated effects, 45
  options, 41–46

Assessment process (cont.)
  other procedures, 45–46
  team planning and availability, 39–40
  variety of approaches, 36–37

Bales, Robert F., 34
Bartelli, Lindo J., 81
Basic geology, importance of 66
Beryllium, as pollutant, 54
Biochemical Oxygen Demand (BOD), 57
Biological web, importance of maintaining, 70–71
Biota, lists of as part of EIS statement, 69–71
Biotics, as significant classification, 178
Borgotta, Edgar F., 34
Bruel and Kjaer Company, 61
Buffington, John V., 13
Bureau of Census, as data source, 39
Burke, Edmund M., 34

California Line Source Model, 55
Calvert Cliffs' case, 21
Camp sites, importance of, 71
Caves, importance of, 66
Cemeteries protection of, 84
Chamber of Commerce, 77
Chemical Oxygen Demand (COD), 57
Chlorides, as parameter of water quality, 57
Ciriacy-Wantrup, S. V., 73
City Council (D.C.), 15
Clawson, Marion, 81
Clean Air Act (1970), 52, 52fn, 54, 54fn

Cliffs, importance of 66
Coliform level as parameter of water quality, 57–58
Colwell, Robert, N., 89
Commercial displacement, 77
Community Development Act (1974), 8*fn*
Community protection, tools for, 80
Computer models, to determine air pollution, 55
Concrete cover, as land destruction, 60
Construction, importance of, 47
Construction alternatives, as option of EIS, 42–46
Construction sites, external noise exposure standards, 51 (T)
Council of Planning Librarians, 46
Council on Environmental Quality (CEQ), 11, 23, 100
  establishment of, 12
  guidelines, 3, 112–46
Courts:
  D. of C. Circuit, 18, 21, 23
  Eighth Circuit, 18, 25
  Fifth Circuit 17, 18, 19, 21
  First Circuit 25
  Fourth Circuit 19
  Ninth Circuit, 14, 15, 18, 19, 21
  Second Circuit 15, 17, 18
  Seventh Circuit, 17
  Tenth Circuit, 15, 17, 18
  Third Circuit, 14

Data collection, 38–39
Data projections, 40
Data sources, for socio-economic assessment, 77
Davies, Gordon S., 46
Decibel scale, 49
Demonstration Cities and Metropolitan Development Act (1966), 10
Department of Agriculture, 38
Department of Housing and Urban Development, 41
Department of Transportation, 40
Design noise level/land use relationships, 50 (T)
Dickert, Thomas G., 3
Dissolved oxygen, as parameter of water quality, 57
"Domino" effects, 44
Dover Book Series, The, 73
Downstream effects, 44
  definition, 37
Draft Environmental Assessment:
  adverse environmental effects, A-39
  contents, A-2-3
  environmental impacts:
    acoustics, A-32
    air quality, A-30-31
    aquatic biology, A-35-36
    climate, A-30
    geology, A-34
    historic/archeological sites, A-38
    socio-economics, A-36-38
    summary of significant effects, A-38
    terrestrial ecology, A-34
    water quality, A-32-34
  environmental inventory:
    acoustics, A-9
    air quality, A-9
    aquatic biology, A-20-23
    climate, A-8-9
    geology, A-18-19
    historic/archeological sites, A-23

Draft Environmental Assessment: environmental inventory (*cont.*)
    socio-economics, A-23
    terrestrial ecology, A-19-20
    water quality, A-11-18
  introduction:
    background, A-7-8
    project surroundings, A-5-7
  list of figures, A-3
  list of tables, A-4
  local use/long term maintenance, A-40
  measures to minimize unavoidable effects, A-39-40
  proposed action and alternatives:
    engineering data, A-24-30
    project description, A-23-24
  resources commitment limitations, A-40
  summary, A-4-5
Draft Environmental Statement:
  basic format, 91–93
  costs and man-hours, 94–95
  submission requirements, 94

Ecological web, changes in, 47
Ecology, definition, 63
Edward E. Johnson, Inc., 62
EIS (*see* Environmental Impact Statement)
Eminent domain laws, 77
Employment, effect on economy, 76
Environment, basic concerns for, 4–5
"Environmental Impact Assessment Methodologies," 46
Environmental Impact Statement (EIS), 37
  (*see also* Assessment Process, Draft Environmental Assessment, Draft Environmental Statement)
  comprehensive qualities of, 1–2
  draft, 91–95
  explanation of, 4–5
  final, 99–100
  mandate for, 7
  problems of interpretation, 5–6
  procedures, 13
  use of the instrument, 8
Environmental inventory, 63–73
  classification check-off outlines, 64–66
Environmental Protection Agency (EPA), 38, 61, 98
  and air pollution, 52
  HIWAY Model, 55
Environmental Research Institute of Michigan (ERIM), 90
Evaluation process (*see* Assessment process)
Excavation, as land destruction, 60
Exchange Bibliography #691, 46
Executive Order 11514 (1970), 31–32
Existing conditions, data collection to establish, 38–39

Federal Aid Highway Act, 25
Federal Air Quality Standards, 52–53 (T)
Federal Highway Administration, 8*fn*, 61
Federal Power Commission, 14
Federal Second Circuit Court of Appeals, 8*fn*
Federal Water Pollution Control Act Amendments (1972), 16, 57
FHWA Acoustics Specifications, 50 (T), 51 (T)
FHWA Policy and Procedure Manual, 49
Field Surveys, use of, 83

Index

Final EIS, 99–100
Fischer, David W., 46
Fishing locations, importance of, 71
Fluorides, as pollutants, 54
Forests, national and state, 68–69

Game lands, importance of, 71
General Radio Corporation, 61
Goldman, Marshall I., 78
Golf courses, importance of, 71
Graphic aids:
  as part of EIS, 44
  use of in presenting impact information, 85–90
Groundwater, 48, 59–60
Gustafson, Joel F., A-35, A-35*fn*

Habitats, loss or change of, 47
"Handbook of Noise Measurement," 61
Hare, Paul A., 34
Heilbroner, Robert L., 78
Herfindahl, O. C., 78
Highway Research Board, 51
Hills, G. A., 81
Historical and cultural resources, 65
  importance of, 82–84
Historic Preservation Act (1966), 83
Historic Sites Act (1935), 83
Hunter, Floyd, 34
Hunting locations, importance of, 71
Hyman, Herbert H. 35

Impacts:
  activities determining, 48
  factors affected negatively, 48–49
Industrial pollutants, 54
Inter-governmental Cooperation Act (1968), 10
Interstate Commerce Commission, 7

Jarrett, Henry, 78
Judson, Sheldon, 73

Keene, John C. 35
Kneese A. P., 78

Lacate, Douglas S., 81
Land conservation projects, importance of, 67
Land destruction, 49, 60–61
Landsberg, H. H., 78
Land use, 79–81
  changes in, 47
  Section 4 (f), 174–75
Lead, as pollutant, 54
Leet, Donald L., 73
Lowman, Paul D., Jr., 90

"Major federal action" interpretations, 14–15
Man-made water resources, importance of, 68
Maps, use of in presenting impact information, 85
Mason, Kenneth G. M., 46
Mathematical evaluation system, 39–40, 178–88
  individual item evaluation sheet, 179–80
  major classifications, 178
  sample evaluation sheets, 182 (T), 183 (T), 186 (T), 187 (T)
  simulated problem, 181–88
McAllister, Donald M., 81

McGauhey, P. H., 61
McHarg, Ian L. 81
McHarg's weighting system, 44
Milbraith, Lester W., 35
Moen, Aaron, N., 73
Multi-discipline group, usefulness of, 39

National Capital Planning Commission, 15
National Environmental Policy Act (1969), 1
  judicial interpretations:
    consultation with other agencies, 22–23
    content of impact statement, 19–20
    Council on Environmental Quality, 23
    delegation, 18–19
    exclusions, 15–16
    innocent bystanders, 19
    list of cases cited, 26–30
    major federal action: definitions and implications, 14–15
    methodology, 21–22
    negative declarations, 17–18
    public participation, 23–24
    quantification, 20–21
    responsibility for preparation, 18
    segmentation, 24–26
    "significantly affecting," 16–17
    timing, 22
  key sections of act, 12–13
  passage of, 6–7
  text of, 104–10
National Environmental Policy Act assessment process, 1–2
National policy, declaration of, 12
National Register of Historic Places, 84
Natural drainage patterns, importance of, 67
Natural impoundments, importance of, 68
Natural resources, 66
  importance of, 72–73
  loss of, 47
  as significant classification, 178
Negative declarations:
  definition, 38
  NEPA and the law, 17–18
  as part of draft statement, 94
Neighborhood structure, changes in, 77
NEPA (*see* National Environmental Policy Act)
Nitrogens, as parameter of water quality, 58
No-action:
  as alternative, 5
  as option of EIS, 41
No-build, as option of EIS, 41–42
Noise pollution (*see also* Acoustics)
  measurement of levels, 49–50
Nonpoint sources, of water pollution, 59

Objectivity, necessity for in preparing EIS, 43
Odum, Eugene P., 73
Office of Management and Budget (OMB), 11
Office of Management Circular A-95, Revised, 148–65
Open space, loss or development of, 47
Outcroppings, value of, 66
Overlays, use of in presenting data 86

Pacific Islands Trust Territory, 15
Particulates, as pollutant, 54
Peterson Series, The, 73
pH, as parameter of water quality, 57
Phosphates, as parameter of water quality, 58

Photomosaics:
  example, 88 (fig.)
  use of, 89
Physical effects, list of, 47
Picnic grounds, importance of, 71
Planning commissions, as data source, 39
Point sources, of water pollution, 59
Pollution:
  acoustics, 49–52
  air quality, 52–56
  groundwater, 59–60
  land destruction, 60–61
  overview, 48–49
  surface water quality, 56–59
Project review, history of, 10–11
Project time schedule, publication of, 33
Prominent elevations, importance of, 66
Public health, impacts on, 77–78
Public hearings:
  basic interest patterns, 98
  preparation and requirements, 96–97
  required by law, 33
Public participation:
  encouragement of, 31–32
  strategies to insure, 32–33
Public service levels, importance of quality of, 76–77
Public services, impacts on, 77

Quarrying, as land destruction, 60

Raw materials, consumption rate of, 72
Recreational facilities, 65
  importance of, 71–73
Redevelopment Land Agency (D.C.), 15
Reforestation projects, importance of, 69
Regional clearinghouses, 11
Remote sensing, use of, 89
Residential displacement, 77
Resorts, importance of, 72
Revelle, R., 78
Review process, importance of, 34
Roadside rests, importance of, 71

Scenic areas, loss or development of, 47
Scherz, James P., 90
Secondary effects, concern with in preparing EIS, 43–44
"Section 4 (f)," 174–75
Senate Document #97, 62
"Significantly affecting," definitions of phrase, 16–17
Socio-economic assessment factors, 75–76
Socio-economic effects, 47
Socio-economics, as significant classification, 178
Soil Conservation Service, 46, 98
Sprawl, problems of, 79–80
State clearinghouses, 11
  national list of, 168–71
State Departments of Commerce and Industry, 39
State Environmental Agencies, as data source, 39
Stern, Arthur C., 61
Stevens, Alan, R., 90
Stewart, Charles L., 81
Strip development, problems of, 79–80
Strip mining, as land destruction, 60
Strong, Ann Louise, 35
Sulphates and sulphuric acid, as pollutants, 54

Sulphur dioxide, as pollutant, 54
Supporting biological web, 65
  importance of, 70–71
Surface water quality, 48, 56–59
  human influences, 58–59
Suspended solids, as parameter of water quality, 58

Tables, as part of EIS, 44
Temperature, as parameter of water quality, 57
Timber forest, destruction of, 60
Todd, David Keith 62
Topographic and scenic items, 64
  importance of, 66–67
Total alkalinity, as parameter of water quality, 57
Total dissolved solids, as parameter of water quality, 58
Tourists' accommodations, importance of, 72
Trade-offs, 5, 72, 77 178
  in draft statement, 93
Traffic study, strategies for air improvement, 55–56

U.S. Army Corps of Engineers, 98, 100
U.S. Congress, 5, 15
U.S. Department of Commerce, National Oceanic and Atmospheric Administration, Environmental Data Service, A-8
U.S. Department of the Interior, 38
  Bureau Sport Fisheries and Wildlife, 69fn
  National Park Service, 46
U.S. Geological Survey, 44, 46
U.S. Supreme Court, 69
U.S. Treasury, 19
U.S. Water Resources Council, 61
U.S. Weather Service, 38
Universities, as data service, 38
Upgrading existing facilities, as option of EIS, 42
"Upstream" effects, 44

Vegetation resources, 64–65
  importance of, 68–69
Vehicular emission, 54 55
Viohl, Richard C., Jr., 46
Vocabulary level, importance of in statement 44

Wagner, Richard H., 73
Warner, Katherine, 35
Water channel changes, importance of, 67
Water quality, parameters for determining, 57–58
Water resources, 64
  importance of, 67–68
Water table, as important measurement, 59
Way, Douglas S., 81
Weaver, Kenneth F., 90
Weighting systems (see also Mathematical evaluation system)
  use of, 44–45
Wetlands, importance of, 68
White water, importance of, 67–68
Wildlife resources, 65
  importance of, 69–70
Wright, Russell, 83fn, 84

Zoning Commission (D.C.), 15